P9-DCD-424

TELL ME EVERYTHING YOU
DON'T REMEMBER

TELL ME
EVERYTHING YOU
DON'T REMEMBER

THE STROKE THAT CHANGED MY LIFE

CHRISTINE HYUNG-OAK LEE

ecco

An Imprint of HarperCollins*Publishers*

This is a work of nonfiction. Some names have
been changed or omitted to protect the privacy and
anonymity of the individuals involved.

HarperCollins books may be purchased for educational, business,
or sales promotional use. For information please e-mail the
Special Markets Department at SPsales@harpercollins.com.

FIRST EDITION

Designed by Suet Yee Chong

Title page image © Potapov Alexander/Shutterstock Inc.

Library of Congress Cataloging-in-Publication
Data has been applied for.

ISBN 978-0-06-242215-6

17 18 19 20 21   RRD   10 9 8 7 6 5 4 3 2 1

For Penelope

TELL ME EVERYTHING YOU
DON'T REMEMBER

1

For thirty-three years I had a hole in my heart and I did not know it.

There was the actual hole in my heart, an undiagnosed birth defect, with which I lived.

And then there was the hole in my heart that I tried to dam up with other people's needs and then filled with resentment. The resentment spilled out as anger, as a need for control, as an obsession with perfection, as an obsession with cleanliness and disinfecting doorknobs and wearing latex examination gloves while typing, as compulsion, as picking the cuticles on my nails and my feet and collecting empty milk bottles and hiding them in all the cabinets of my kitchen and under the bathroom sinks until my husband found them months later and as he threw them away I wept for the lost bottles, even though my sadness was not about the lost bottles but about something unfulfilled that I had yet to identify or acknowledge.

There was the hole in my heart that made it hard for me to breathe.

There was the hole in my heart that made it impossible for me to be whole.

And then I had a stroke.

———

On December 31, 2006, the sky was blue, there was snow on the ground, and I was in the parking lot of a hardware store with my husband. I was thirty-three years old. We were doing the exact same thing we'd done the previous New Year's: spending the holidays in Tahoe, going to Ernie's Coffee Shop for a late breakfast of eggs and toast, playing board games, and drinking hot chocolate in the evenings.

Everything appeared normal, but unbeknownst to me, my life was on a knife's edge; my body, mind, and relationships would undergo immense changes beginning that day, that new year to come, that lifetime ago.

I had woken up with a headache. I complained of great pain. I curled up in bed and squeezed my eyes tight, wishing myself back to sleep. The thought of food made me nauseous. Right then a clot had made its way through my body into my heart. If there had not been a hole in my heart, all my blood and the clot with it would have traveled into my lungs. There my blood would have filled with oxygen; there the clot would have been filtered out, a non-incident. But because there was a hole in my heart, the clot skipped my lungs and went straight up into my brain, traveling until it could not travel any further. There it stuck. There it obstructed my brain from receiving oxygen. There it killed a part of my brain.

We thought it was another one of my migraines. Maybe I should drive a nail into my palm, I thought—maybe then I would be distracted from the unrelenting, stabbing pain in my head. Maybe I just needed some fresh air. Maybe Adam and I could run an errand and drive with the windows down, the frosty, pine-scented wind crashing into our faces. The thought of freezing-cold air lured me outside. I wanted to submerge my head in a snowdrift, to encase my head in ice.

That drive was a waterfall of sensation during which I lost my grip on meaning. I heard things but could not take in the sounds and imbue them with any significance. Somewhere between noise and meaning, vision and meaning, touch and meaning, communication broke down. Numbers became squiggles, colors lost their names, and music had no melody. There was a cascade of sensory input—triangles and sky and gravel sound and music on the radio and the feeling of rough cloth near my hands. I could not make sense of any of it; I did not know that the small triangles were trees, the larger ones mountains, the sounds were tires crunching snow and Snow Patrol playing on the radio, the jacket was Gore-Tex, my wrists were the things attached to the things called my hands. They were colors and shapes and sound and touch and sensation, and my brain was no longer sorting these things out.

But when we parked and I stepped out and saw the red snowblowers in the parking lot rotated ninety degrees and doubled, I finally had a complete thought. I was able to comprehend what I was seeing before me: red snowblowers. But sideways. Strange.

In fact, my whole world had rotated ninety degrees.

The sky was to my left. The ground to my right. I was out of sync.

I turned my head, in hopes that I could right the world.

This is not normal; this is beautiful, I thought. But I was dizzy, as if on a boat. And my head hurt.

"I need to sit down," I managed to say. I had not yet lost my words in the middle of this parking lot. "You go inside, and I'll sit here."

I let go of his hand.

Adam said to sit on the curb outside the store, not in the parking lot. That he would be right back.

It wasn't easy to sit down on the curb; I had to allow my body to lower itself, even though my brain told me the sidewalk was standing straight up, and I was about to sit on the sky. My world had literally turned sideways. But I did not think I was dying, even though at that moment I might have been staring at the gateway to death. Why is everything so strange? I thought. What is happening? When will everything go back to normal?

Adam disappeared and came out shortly after, empty-handed. "Let's head back," he said. "There's no way I can buy HVAC filters while you're out here. Something's wrong."

And eventually my thoughts subsided. All of them. My brain went . . . quiet. Dark. As much as I try, years later, I cannot remember that ride back to the house. Part of me was choking, deprived of oxygen—dying.

When we got back to the house, I was so tired. It was a depleted sort of exhaustion— every sound felt like a shriek, I could feel the weight of my eyelids when I blinked, and I couldn't even eat, because chewing felt like doing chin-ups. My body screamed for rest, for sleep. I would not be this exhausted again until giving birth to my daughter seven years later.

So I slept.

Sleeping is not recommended immediately after or during a stroke. You should have come into the ER immediately, my neurologist would tell me days later. But I couldn't help myself at the time. I would have taken a nap even if I had been advised not to. There was nothing I wanted to do more.

I dreamt about getting lost in the snowy mountains. I dreamt that my friend, let's call him Mr. Paddington, my husband, and I had started a hike as a group, but I got separated from them and had to continue trekking alone. I dreamt about snow falling. I dreamt about walking along a frozen Alpine lake. I dreamt about losing my shoes. I dreamt about losing my voice. No matter how much I screamed for help, no one came. No one could hear me. Until I realized I was making no sound. I was not alarmed. What was happening was just happening, and I told myself to keep walking forward.

When I woke up hours later, I really believed I was in those mountains hiking—that it had not been a dream. And indeed, I really had lost my voice. I had lost my words. I was unable to say, I am trapped in my brain. That my memories were mixing with imagination.

The thing is, I'd lost my voice in so many ways already, before the stroke even occurred. I had been unable to say, I am trapped in my life. That my obligations were interfering with my personal dreams. I made up rules and stuck to them because that was safe. It was, however, not brave.

This is what I blogged on the evening of my stroke after awakening, in an attempt to communicate what I was experiencing:

> I am feeling strange. My brain is in a weird state right now—a combination of short brain games and lack of memory. While taking on the concept of a brain game earlier today, I suffered a memory overhaul. Now I can't say what I want to say or remember what I want to remember. It's just a weird situation.

I had aphasia. I had become a writer who could not use the correct words. Years later, I understand what the "short brain games" were—but I'm surprised that I'd phrased my experience in that way. I remember being so certain that my words were correct. I thought I was communicating just fine.

Just seventeen hours earlier, pre-stroke, I'd written the following in my journal: "So this is how it feels to hole up somewhere; the snow has come on and off this week, the chilly air outside has the snap of a crisp spring peapod, and all is peaceful. There is no external stimulation; my life has turned inward this week. Reading books."

That was before the aphasia. But accurately enough, my life did turn inward. My life would be inward for a very long time. Before the stroke, things were peaceful. I was reading books. I was reading Kurt Vonnegut's *Slaughterhouse-Five*. I started reading it right before my stroke, and I continued to read it in the days following. But I was reading the same page, over and over. The first page. The first paragraph. I did not know this until later.

Our best friend, Mr. Paddington, arrived for New Year's Eve, excited to join us, and all I could do was smile and say a few words. Just a few words. In the hubbub, I was silent. I am never silent. I also never nap. Not until that day.

"Hi, I'm having a brain drain," I said. I watched myself struggle. Underneath what felt like one hundred down blankets, what was left of my pre-stroke self said, That is not what I meant to say. Something is wrong.

I need help, I wanted to say.

But no one, not even I, could hear or understand.

We went out for a New Year's Eve fondue dinner at the Swiss Chalet, which is a Bavarian-themed restaurant that looks like it belongs at Disneyland adjacent to the Matterhorn ride. We dipped bread into cheese and drank beer and wine, and I remember conversation with which I could not keep up. I was a bicycle trying to merge onto a freeway of speeding cars. Every time I thought of something to add to the conversation, I forgot what it was I wanted to say, and the conversation progressed without me. My mouth opened and closed like this many times throughout the night. Like a fish gulping water.

My world that day narrowed to basic inputs, without processing understanding or meaning. Whatever I saw had no meaning attached to it, as if I'd just arrived on this planet. I felt the sofa underneath me, the velvety Naugahyde that reminded me of my dachshunds' fur. The cold air that stung my face. Conversation that eventually devolved into a din. The unbending metal of what was called a spoon in my mouth. I could tell that what was put in front of me was food, but I could not figure out if it was food I liked or abhorred. It was just food. To put in my mouth. To chew. To swallow.

This is what you're supposed to do during a stroke: you are supposed to see if you can smile, to figure out if you are confused, to see if you can think of words. You are supposed to head straight to a hospital, so they can dissolve the clot with expediency.

The American Stroke Association uses the mnemonic device FAST for help in recognizing the warning signs of a stroke:

FACE DROOPING: Ask the person to smile. Does one side of the face droop? Mine did not.

ARM WEAKNESS: Ask the person to raise both arms. Does one arm drift downward? Mine did not.

SPEECH DIFFICULTY: Ask the person to repeat a simple phrase. Is speech slurred or strange? My speech was not slurred. What I said was indeed strange, but I did not think it was. I could not diagnose myself.

TIME TO CALL 911: If you observe any of these signs, call 911 immediately. I did not.

I did not have all of these warning signs. In fact, I did not have most of them. My brain was dying, but I did not know it. And I had grown to be stoic in the face of pain. So I did not head to a hospital. The clot was stuck. I was stuck.

Instead, we bided our time.

There are pictures from that day. In the pictures I am sitting on the couch, smiling, rested after my nap, cuddled up next to Adam. Mr. Paddington is also on that sectional, reading magazines. As if nothing of significance was happening. As if I wasn't having a stroke. I looked completely normal even though I was technically, literally dying inside.

People have asked if anyone around me could tell I was having a stroke.

"Weren't you acting weird?"

And my husband's mouth would then turn into a thin line. My friend who joined us on New Year's would then lower his eyes. I was acting weird, yes. But I looked fine on the outside. And it was New Year's Eve. Adam and Mr. Paddington were drunk and jolly and distracted. I was not talking. They thought that was odd but not cause for huge concern. They thought that perhaps I too was drunk.

By the next morning, I could speak a few words, albeit some strange ones.

"Good morning. Let's have breakfast. I've lost my data, all I have is its footprint."

My words did not come out slurred, and my face did not look like melted wax. I could smile. I was thirty-three; I could have been drunk, but stroking out? No. The idea of stroke was so remote to us, it did not even feel like a possibility. So we waited for the symptoms to subside.

Weirdest migraine ever, we said. We had breakfast. I know I had two fried eggs, because that is what I always have when we have a hot breakfast.

Later, when I went to the hospital, even the doctors did not at first think I'd had a stroke. In the transcripts of my medical files, I now read:

> The patient is a 33-year-old-female who presents to the hospital with complaints of memory impairment, diplopia, inability to write and sensation of dizziness without imbalance concerning for CNS vasculitis. Migraine is unlikely as this is atypical for her typical migraines. *Ischemic stroke from cardiovascular disease is also quite unlikely.*

Sometimes I wonder. What if I'd died that day instead of having survived, if the clot had lodged for a moment longer and sent me into a permanent coma? If I'd never woken up from my nap? If instead of having gotten up from the sidewalk, I had collapsed? If the last thing I saw was the world tilted on its side? If the last thing I thought had been, When will everything go back to normal?

My life would have been just a chapter or two. A novella instead of a novel. I would not have had my daughter. I would not have lived a life in New York City and pursued writing full-time. I would not have learned what it was like to breathe again. Likewise, if Kurt Vonnegut had not survived Dresden, there would not have been *Slaughterhouse-Five*. If I'd died, I would have had a decent life. I was not caught in a firestorm. I was not held a prisoner of war. But surviving my stroke pushed me to believe that my life could be better. That I had a second chance

at something I could not yet see. That I could—that I *had to*—
live a better life, one based on personal priorities, one centered
on pursuing my dreams, one without regret.

The stroke pushed on the weakest and most untested seams
of my psyche. Where before I could never feel vulnerable—I
wouldn't let myself—I was now vulnerable all the time.
Where before I could not ask for help, I needed to ask for assistance to get through the day. Where before I could not allow
myself to feel sad, I lost the ability to dam up my emotions.
Where before I always planned in the interest of being in
control, it was now impossible to do so. A part of myself really
had died.

I was a body more than anything else, because my mind
was on break. My mind was at peace. All the chatter in my
head—*What should I make for dinner? I need to go grocery shopping, but maybe I should go do that after my doctor's appointment, because then the groceries will stay cold instead of sit in a
car, and the store is on the way home, and so I won't have to back-
track, and wait, do I have enough gas? When was the last time I
put gas in the car? What if I run out of gas?*—all that chatter was
absent. All the burden of planning, all the anticipation, all the
worrying and fretting, the burden of thought itself, was gone
those first few weeks of recovery.

If the stroke had been more severe—if the clot had stayed
in my thalamus and choked it further, kept it from oxygen
any longer, I would have ended up in a coma. I would really
have just been a body. The body would have won in its battle
against my mind.

And if I'd been in a permanent coma, I would want to have
been released from this life.

Once, years before my stroke, I brought home a personal directive form from my doctor. I'd signed it. I did not want to be put on life support, should I need assistance to breathe long-term.

But Adam said he would keep me on life support.

"Even if I signed something that said I didn't want to be on life support? Even if I would be suffering? Even if there was no hope I'd be the same again?"

"You can't kill yourself," he said.

"But it's my decision to make, regardless. It's my life. And it's my death," I said.

"It's not your decision. It's mine."

"Why?"

He explained. "Because I'm the one who has to live with your decision. The person who survives is the one you have to think about—the person who makes that call."

The idea of asking for permission to die the way I wanted to die was unfathomable to me.

And then he made me rip up the personal directive I'd signed. It didn't seem a significant interaction at the time. I shrugged as I threw the pieces into the garbage. He wanted me to live, I thought. He cherished me that much.

Even if I remained on life support, I figured, I would eventually die. We all do. Or at the very least, I would no longer be an active participant in my own life. I would be asleep forever. A kind of death.

In a sense, I was already not living my own life. I thought I was, because I had a career and I paid my own bills. But I was living Adam's life, cheerleading him and not myself. It was not his fault that I did this. We both felt comfortable with that dynamic. It was what was familiar. And for many years we were very functional this way.

The last news item I remember reading before my stroke was a story about two people named James and Kati Kim. On Thanksgiving weekend they'd driven their Saab into the Oregon wilderness in error and gotten stranded in the snow with their two small children, Penelope and Sabine, one of whom was only an infant. They burned paper and tires for heat. Kati Kim breastfed both her daughters to keep them alive. James Kim ultimately walked away for help, promising to return by the end of the day.

Meanwhile, James's family struggled to find them. His sister besieged the wireless phone company to track the pings of their cell phones. His father hired private helicopters to supplement the search effort from the air.

Kati started walking with her kids the morning after James's disappearance, and after nine days trapped in the wilderness she was spotted by a rescue helicopter.

But James Kim never returned. He was found face up in Big Windy Creek's freezing water two days later.

It was the biggest search and rescue effort in Oregon history.

I never forgot this story. I'm sure it informed the dream I had in which I walked through snow, looking for rescue. This dream was the setting in which my brain died. That part of my brain is out there facedown in the water, having perished. I did not come back the person I was, either. I was separated from myself, and I grieved for that part of my brain. I was heartbroken without it. I didn't think I could go on.

But then I did go on.

When I finally went to the hospital, the doctors conducted a memory test, using the word *penny*, the name I would later give my daughter. The same name as James Kim's daughter. Mr. Paddington was there that day, as he is with me now. I lost access to my photographic memory, without which I lost faith in my own intelligence.

I was reading that Kurt Vonnegut book over and over those first few days, not knowing I was doing so. My brain was not functioning well enough to remember more than a paragraph at a time, and I would have to start over after a few lines. I did not realize I was reading the first paragraph of *Slaughterhouse-Five* repeatedly.

"All this happened, more or less," I read. And yet I did not realize what was happening. *"The war parts, anyway, are pretty much true."* The childhood parts are pretty much true. I remember what I can remember. Piece together what I can. It took me years to write about my stroke, like it took Vonnegut years to capture Dresden's firebombing in writing.

The more I think about that day, the more I realize that it was an unintentional reboot of my life, a template for a new start.

What I lost had to be lost.

2

The first moments of the stroke shook the foundations of my brain. Nothing worked as it should in that first hour—I couldn't see properly, my balance was off, and I had an immense headache. I could barely form coherent sentences. But when I woke up from my nap, I had already begun to heal. My vision went back to normal. Objects undoubled. My headache was gone. I could write some words, even if they were nonsense. But I still couldn't remember or assemble ideas in a comprehensible manner, or retrieve proper words in the proper order; I could not yet communicate. After the fondue dinner I went to sleep early, leaving Adam and Mr. Paddington to watch the ball drop on television.

Inside my brain, all the files had come out of their boxes and all the papers had come out of the file folders.

The next morning there was activity all around me—breakfast being made, groggy chatter, *Harold & Kumar Go to White Castle* blaring on the television screen, and I felt like

I knew how to participate, but when I tried to think about where eggs might be stored, I didn't know. I tried to say *egg*, but I didn't know what the word for "egg" might be. I tried to describe eggs, but it didn't come out "those elliptical shell things"—it came out "shell bells, where are the shell bells?" By then, everyone had finished making the eggs and moved on to making pancakes. I did not know the word for "pancakes" either, and I did not know how I wanted my shell bells prepared because I forgot I always wanted my eggs fried over easy with the yolks runny and the whites firm, and I didn't know what preparation meant. I chewed my cooked shell bells. I saw the batter on the griddle, and I smelled it cooking and rising and bubbling, and they were talking, and again I wanted to participate, but I did not know how to say *pancakes* or the word for stirring pancakes. So I said, "shell bells," because I knew shell bells go in pancake batter. I figured that much out. And that the shell bells would help. They would help. Help. Help.

And after a while, I gave up. I was so tired. I didn't know how to explain shell bells to anyone who asked. And I didn't know how to say "Help." But I did know how to say yes. And so I kept saying yes. And smiling. Yes. Yes. Yes. It made everyone happier. So I was happy. I smiled. Hi. Hi. Hi. How are you? I felt connected in that way. I kept being pleasant. I liked the smiles. They meant something to me. And with smiles came zero questions. With zero questions, I felt less helpless.

"How are you feeling, Christine?"

"I am fine." I didn't know what else to say. I was surprised that I knew those words, "I am fine."

I now understand why I did not know the words for "I

need help." I was not in the habit of asking for help. It had become a habit for me to say I was fine. It bothered no one when
I said I was fine. It was thus easier for my brain to shoot out
that automatic verbal response.

Help, on the other hand, was a new concept. My brain could
not build new things. It was busy repairing the old things. Help
was difficult. Help was complicated. So, I am fine, I said.

I chewed my pancakes. They did not taste sweet or savory
or vanilla, and I did not register pillowy cake or caramelized
crust. The food did not spark joy in my head, and this was sur-
prising, because food usually sets off a cascade of emotions
and memories and pleasure for me, and this in turn causes
me to crave and eat and oftentimes overeat and then gain
weight, but there was nothing like this that day. The shell bell
cakes were simply going into my mouth.

I didn't know how to say that everything tasted bland. That
a bowl of rice tasted the same as a fried egg as a pancake. That
I could see they were different foods, but the memories associated with each thing were gone. That the flavors were no longer
meaningful or unique. That the flavors no longer connected to
my emotional center—that I no longer had favorite foods. That
I no longer had food preferences. I no longer remembered that
my father's only cooking skill, besides making instant ramen,
was knowing how to fry an egg. How he would fry up eggs
for us every day when my mother was in the hospital for her
kidney-stone operation. How because he liked eggs well done,
our eggs, too, were served with the yolks dense and yellow-
green. How in spite of my childhood eggs, I now like my eggs
over easy, the yolk flooding onto the plate like a rich sauce.

That I had no appetite, or thirst.

That I had no desire, really.

I was an empty shell bell.

———

Three hours later I say it. "Egg!"

"What?"

"Egg!"

"What?"

"Shell bells are eggs. Eggs are shell bells."

"Okay, Christine."

I was spaced out. So spaced out. My brain was in space.

I showered. The water was wet, but I no longer even knew that word. I was going through the motions I'd gone through nearly every morning of my life. Turned the metal knob thing. Took off the things on my body. Hung the rectangular soft cloth by the shower door. Stepped in. Stood under the wet stuff. Washed the stuff on my head called hair with slippery stuff called shampoo. Applied the sudsy stuff on my body, arms first, then neck, torso, and legs. Rinsed the sudsy stuff off. Stepped out. Dried the wet off my body with the soft, rectangular cloth.

The music was on. And it was noise. I did not care for it. It had no meaning. It had no memories. I was only listening to noise. It was different from conversation, but the same in that I could not comprehend it. I was not aware of this lack of meaning. It just was.

———

This is what I wrote on my blog months later about that first day:

> I thought that a chunk of my brain was off somewhere, having
> the time of its life, without me. I was aware that my memory
> was gone; I would try to recall things, but when I would direct
> my mind to that very memory or task, I would be confronted
> with a dusty footprint of knowledge that was no longer avail-
> able. My warehouse felt empty. Even though it was not. I had a
> catalogue, but I didn't know where anything was. Every single
> thing felt like a fleeting thought—that feeling where you wake
> up from sleep and know you dreamt, and have an impression
> of what it is you dreamt, but try as you might, you cannot re-
> member the dream.

Every single moment was like this. A haze. A dream world
of intangibles. Of reaching for and grabbing air.

Normally, I would be very frustrated by such a state. Normally, I am frustrated when I cannot remember something,
even something minor, and know that I should.

I was not normal. I accepted this malfunction as fact. And
slid into a comfortable haze.

I picked up a book. Kurt Vonnegut's *Slaughterhouse-Five*.
And began to read.

*"All this happened. The war parts, anyway, are pretty much
true."* Back to page 1.

For a couple of days I believed the stuff of my dreams really
happened, only to have Adam tell me a few days later that they

never did. This went on for months afterward. I wasn't hiking in a forest. I was not sailing the ocean. I was not flying. I had to throw all of my dreams out of my mind to stay in reality.

I lost all the dreams, the soft edges of those memories melting and then shelving themselves in some part of my brain I could no longer reach.

And I was no longer dreaming anyway.

In place of these dreams? Some fantastic impressions: the brief memories of a storm, of a huge fight in my brain. Of characters.

I was trying to pick up the pieces.

I kept reading Vonnegut. So it goes.

Two days after the stroke, two days into 2007, we returned home to Berkeley.

I was drowsy the entirety of the 180-mile ride. I don't know how it is that by this point we didn't know something was seriously wrong. Maybe I napped the entire way, evading detection once again.

Years later, my husband said that he should have noticed something wrong by then.

By nature, during our drives, Adam and I never did talk much. We expected and enjoyed miles of comfortable silence. And at that time, my new cognitive functions weren't yet tested. We had also done that drive many times, preoccupied with thoughts of work the next day and errands to run once home. If there was talk, it was usually banal. So the silence was infinitely ordinary.

In hindsight, I also have to take into consideration the person I used to be—someone who did not notice her body, who did not acknowledge pain, who never asked for help. My father valued stoicism in the face of pain and strength at all costs; this was a man who once lay in bed with severe pain refusing to go to the hospital, and then when he finally relented days later, insisted on driving himself, until he realized he was unable to drive because of the pain. That was the first time my mother ever drove him around as a passenger in a car. He went straight from the ER into surgery, where they discovered a perforated small intestine. Ultimately, the doctors had to cut out two feet of torn intestinal tract. He still brags about toughing it out. That he tried to drive himself. That he stuck around at home for three days, steadily going septic.

Decades later my mother would drive him for the second time ever, after he suffered an enormous stroke. She would drive him home from an inpatient rehabilitation center.

My mother was an ICU nurse who witnessed her own share of tragedy. Sometimes I could tell how her day went when she volunteered small pieces of information, always oblique. On those days she would blurt, "Christine, don't ever tattoo your eyebrows or lips. It looks awful when you die. My patient had eyebrow and lip-liner tattoos. It was not good."

I wanted to ask how her patient died. Whether or not she was saddened. Why this particular patient and not another. But I'd asked these questions before, and I always got the same answer—something along the lines of "Death happens. But I'm still alive."

Death happens. I am still alive.

There is a running joke among children of doctors and nurses—that you can always tell who we are, because we are the children with runny noses and tearing eyes. Because nurses think in terms of life or death. And a cold is not death. The flu is, to an intensive care unit nurse, not immediate death. I did not know I had chicken pox until after the fact, when my brother came down with it. Until we detected a pattern.

So it went with my stroke. Death happens. I was still alive.

———

After we got home to Berkeley, Adam and I piled straight into bed, comforted by city sounds in the darkness—the BART train rushing through the hillside below us, the freeway a low hum beyond that. When I woke up the next morning, I remained disoriented.

"I still don't feel well," I told Adam. "I'm staying home from work."

"Okay."

I drove to the store to do some grocery shopping—our fridge was empty.

I went inside Andronico's grocery store and browsed the aisles, a blur of colors and letters and shapes. What was it we needed? I wondered. I could not figure out how the pieces fit together—that I would need onions because we used onions for everything we cooked, that I would need bread for sandwiches, and meat for a possible entrée. Everything had become shapes and colors and textures that I could not comprehend. That mushy package was a fleshy pink rectangle and not ground beef. The countless cans of soup and vegetables were mere metal cylinders composing a kaleidoscope of chaos.

From that store I emerged with one thing: a jar of Muir Glen spaghetti sauce. I grabbed it because it felt familiar to me, and for a split second I could identify the label. If it was something I could understand, it must be something I needed. I did not, by the way, need spaghetti sauce.

I still do not remember how it is I paid, whether by cash or by debit or credit card. I do not remember swiping a card or handing over bills. I just assume one of those things happened, because I do not remember an employee running after me or the cops being called. I just remember blinking in the cold winter sun at the cars in the parking lot. Clutching a jar of spaghetti sauce.

And wondering how to get home. I did not know how to get home.

It did not occur to me to think about how I'd gotten there in the first place. How *had* I gotten there? I was lost in time. I was lost like a dog who runs miles from home and then can't figure out the way back. Except I should have known the road home. I was not a dog.

I hugged the jar of spaghetti sauce in that parking lot. I bit my lip. I was helpless. If someone had asked me my name at that moment, I might not have known. No one talked to me, though. I just stood there with my jar.

And where was my car? Which car was my car? In my hand was something I felt would help me find the car. It was my key. I pushed the button, and my car chirped. Somewhere. I turned my head toward the noise.

I was in an entirely new world, as if I'd walked from one

dimension into another in which everything looked the same but nothing was the same at all.

But really, it was I that had changed.

It did not occur to me to use my phone. It did not occur to me to use the navigation system in my car. I did not know my address, anyway. My home address was permanently encoded in the GPS, but I would not have been able to operate the system, even though I used it all the time. The knob was not a knob, merely a circle, and the map was only squares and squiggles, and I could not put together that these things would help me get home.

I got into the car and started driving. If I just drive, I thought, I will somehow get home.

I had faith that this method would take me home. That I would take myself home.

Each time I thought about whether I needed to make a left turn or right turn or stop or go, I felt lost. I had no idea. The stop signs were red things. When I stared at them, I could not figure out what the white shapes on the red background meant to say. But somehow I stopped at every one, anyway. I can't promise that I did. But I have no tickets. There are no pedestrian complaints. The car was intact. I drove home on automatic pilot, in the same way I knew how to brush my teeth each morning—outer top teeth, inner top teeth, then outer bottom teeth, inner bottom teeth, tongue, then rinse. Each time I looked around to try to check in, I recognized landmarks—a tree or a house or a store. I did not know where they were or exactly what they were, but I knew I was getting closer to home. I pressed on without thinking, because thinking got in the way; to address my conscious-

ness and try to figure out what steps to take next was para-
lyzing.

Intuition carried me when logic and memory failed.

And there it was—our house.

I looked at the key in my hand and wondered where it would
go. I stopped thinking about what a key was and how it would
work and where it should go and released the key from my conscious mind, and before I knew it I'd opened the front door to
set down my jar of spaghetti sauce on the counter. Like magic.

I was home. It was a house I knew very well, a place in
which I could walk around with my eyes closed. I didn't have
to think to navigate, and I didn't have to wonder where everything was. My body had tracked its walls and windows for so
many years that this place was where I felt most comfortable
at that moment. Safety.

But the jar of sauce disturbed my sense of safety; it was a
symptom. But of what? And our fridge was still empty. Had I
locked the car? How did one lock a car?

And then I thought, I need to get to a hospital.

I picked up the phone and asked myself, "What is the
phone number for 911?"

I looked at the numeric keypad but could not figure out
what number each shape represented.

And what was the number for 911?

I thought perhaps I should try calling Adam. Yet I could not
remember his phone number. It did not occur to me to look for
it in the contacts list on my BlackBerry. I would not have known
how to do that. My problem-solving capabilities had left me.

I do not know for how long I sat there wondering how to
dial 911, but I finally decided I would mash a bunch of numbers

on the keypad and talk to whomever it was I ended up reaching for assistance. I did not think about the fact that I did not know where I lived, and thus could not attain help, but I punched in a random set of numbers anyway.

"Hello," a man said.

"Hi!" I said.

"Hi," he said.

"Who is this?" I asked.

"This is Adam," replied the man.

"Oh! I have been trying to reach you! I forgot your phone number, and I didn't know how to get ahold of you! I called this phone number, because it was in my fingers."

"I'm coming right home," said Adam.

———

We went to the nearest emergency room, two miles away.

"Something is wrong with my wife," said Adam.

The receptionist told us to wait.

The waiting room was dark, even though every fluorescent light was lit. There were no windows. It felt like a basement. I sat, thankful for the darkness and lack of input. I held *Slaughterhouse-Five* in my hands. It might be a long wait, I thought. I began to read.

"My wife can't remember anything," said Adam. "Why can't you see my wife right away?"

Somewhere in the corner a man was moaning, doubled over.

"I've been waiting over two hours!" a woman shouted from another corner. "They say they've got a gunshot wound back there!"

"But my wife—she barely knows her own name. She doesn't know what a number is. What if she's dying?"

"We're all dying in here!"

We waited. My husband paced. "It took forever for us to decide to get here—why are we waiting now? We're leaving. This is unacceptable."

I kept reading. *So it goes.*

And then my husband strode toward me, took my hand, and led me out of the emergency room.

"We're going to another hospital," he said.

We drove to the newest, biggest hospital, four cities over and fifteen miles away. The emergency room there had windows and smelled like a living room; they called me within five minutes. I know this, because they gave me forms to fill out, and then called my name as soon as we turned to sit down.

I went through the double doors to another space filled with nurses and doctors and doors to treatment rooms and machines. In our treatment room, lying on the bed, I did not know how to fill out the forms. There was the word *name.* I recognized it at that moment. I knew my name. That was it. I wrote my name, and when I looked at what I had scribbled, it looked correct. But what were the other boxes? What was my address? What was my phone number? I turned to Adam and said, "Please do this for me."

In the back of the ER, the doctors gave me a CT scan to get an idea of what was going on in my head. I do not remember having the CT scan. But the hospital records said I did. Adam said I did. It happened quickly.

My dreams mixed with reality. I was in the ER, but where

had I been before? How long did that CT scan take? Why couldn't I remember it?

"All this happened," I read.

Along with Vonnegut, I'd taken my Moleskine journal, a perennial resident of my handbag, with me to the hospital.

In its squared pages I wrote, "I've had my CT and the doctor wants me to have an MRI first thing in the morning so he's admitting me. That's five hours condensed into one sentence." I added, "Already, I've sat here and watched a family grieve over their just-dead grandmother."

Because of my journal entry, I now remember. One room over in the ER, someone died. I remember the rising cadence of activity. Running footsteps. And then I remember sudden silence.

I remember a doctor speaking in a hushed monotone, down the hallway. There was a parade of weeping people. I think they were polite people; their crying was measured and bleating. As if they didn't want their cries to pierce the hopes of everyone else in the ER.

"All this happened."

In the hospital dictation notes, my neurologist wrote:

The patient is a 33-year-old woman who 4 days ago had sudden onset of vertigo described as like everything tilting. This was also associated diplopia and speech difficulty. The vertigo and visual disturbance resolved between 5 and 15 minutes after which the patient felt sleepy. On awakening, she found she still had speech difficulty and memory problems. These have been gradually improving, but caused her to come to the Emergency Room last night. She has a mild expressive aphasia. She is able to name only 14 animals in 60 seconds.

There's a dark spot on the CT. "We think you have vasculitis," said the doctor. In my files, I later read the hospitalist's
dictation: *patient has focal low attenuating area in anterior left
thalamus.*

In other words, there was a scar in my brain.

At the time I said, "Okay."

He continued. "We think it's vasculitis, and we need to admit you for more tests."

My husband joked, "We need Dr. House."

I had what felt like ten doctors in the ER: a hospitalist, the
attending physician on call, the neurologist, another backup
neurologist, a cardiologist. I cannot remember what the
others did.

What my neurologist, Dr. Brad Volpi, did not say, I later
read in my dictation notes:

> Impression: subacute left thalamic stroke in a young adult,
> essentially healthy. Vasculitis and dissection need to be
> ruled out, also a new neoplasm.

Dr. Volpi told me it was vasculitis or a torn blood vessel.
He also suspected a brain tumor. He did not tell me he suspected a stroke, even though he'd written it down as a possibility. He also did not tell me he was considering a brain
tumor. He would not tell me these more serious prognoses
until he was sure.

"All this happened, more or less."

We waited for a bed in the hospital. They did not have a
bed in the stroke care unit, which was called the DCU or definitive care unit then, nowadays called the PCU, or progressive

care unit. So they admitted me and placed me in the maternity ward, in postpartum. I was the only patient in that desolate, darkened wing. Were no babies being born that day?

In the room there was a bathroom. In the bathroom there was a shower. There was a bed for a guest. There was a bed for me. I changed into a gown and crawled into the bed. I was tired, as I'd been all day. I curled up.

The nurses attached me to wires and continued to interview me for intake. "What medications are you currently taking? Do you have any other health conditions? Do you have a history of high blood pressure?"

All these questions Adam answered. I was too exhausted to speak. I had lost all my words again.

Their fingers tickled when they stuck the EKG stickers on my breasts.

I opened my book. *"And so on."* I kept reading.

The next morning I was wheeled to the MRI room. I lay flat on my bed and stared at the ceilings the entire way. Some hallways were cold and gray, others warm and filled with yellow morning light. The elevator's ceiling was metal. The MRI room was dark.

Many people hate MRIs—it is not a physically painful experience, but it can feel claustrophobic. I liked the MRI. It was certainly more comfortable and less intrusive than the transesophageal echocardiogram (TEE), more peaceful than a chest ultrasound.

I took off my wedding ring and my necklace, all my metal

jewelry, and lay down on what was essentially a non-metal slab, which then slid into a narrow space that reminded me of an opaque test tube. Or like one of those Japanese capsule hotels. My hospital gown hung from my body, and as a gust of air conditioning wafted between my skin and cotton, I briefly wondered if the worn fabric, blue and dotted with diamond-shaped hashtags, was see-through. But by that point, so many people had seen my body, I no longer cared. And even if I had cared, I wouldn't have remembered to keep worrying. Worrying is an exercise in memory.

It was peaceful inside the tube. I had been craving a small space. The world had been too big since my stroke. There were thumping noises. Clicks. Grinding sounds. I closed my eyes and imagined a beach, ocean waves crashing, a distant horizon. I wondered what kind of imagery that would produce on the MRI—if the radiologist would see that I was thinking of water. I wondered if my brain would light up like a beach sunset, the western sky peach and the eastern sky doused in shadow.

This was the first of five MRIs I would have that year.

I was wheeled back to my new room. The hospital's ceilings were very clean.

They transferred me to the definitive care unit, where a bed had opened up overnight. The unit was bustling. The lights in the hallways were bright. The patients sounded like old people, the groans coming from well-traveled and graveled voices. The nurses' faces registered surprise when they saw me. I was surprised and delighted when I found that the toilet pivoted out from under the sink. "Rad," I said. "Rad."

While I was in the MRI, Adam had driven home to fetch my pajamas. The nurses from the night previous said that I should make myself comfortable, including donning my own pajamas if I wished. So I changed into my Paul Frank pajama bottoms and a black T-shirt with a drawing of a bunny and a speech balloon that said, WHO'S YOUR BUNNY?

I got into bed. I took out my book. *"All this happened."*

And then Dr. Volpi came to me with the MRI results. "Hi, Christine. We've discovered that you've had a stroke."

Okay.

I couldn't believe what I'd heard. Adam couldn't believe what he'd heard. Dr. Volpi was surprised, too. Adam asked him to repeat himself.

But then I couldn't remember what had been said. I hit Rewind.

"What did they think I had, Adam?"

"Vasculitis."

"And what did I have?"

"A left thalamic stroke."

"Okay. But . . . what did they think I had?"

I'd had a stroke. I looked down at my book. It was then that I realized I had been reading the same page for days. The same three opening paragraphs, the one page thumbed, the rest of the book undefiled.

> *All this happened, more or less. . . . I've changed all the names.*
> *. . . There must be tons of human bone meal in the ground.*
> *. . . So it goes.*

The planes found their way to Dresden.

3

"What was it they thought I had?"

"Vasculitis."

"Oh. And what did I actually have?"

"A stroke."

"And what is it they thought I had?"

The morning after admission, after my MRI, the doctors told me I'd had a left thalamic stroke. It was at this point I realized I could not remember what I'd been reading. Or what the book was about, even. It was just snippets of lines that faded minutes after being read, forcing me to read them over again and again, without even being aware of doing so. Once my mind absorbed the fact that my brain was not functioning as it should, I put down *Slaughterhouse-Five*.

I thought it would be easy for me to write about the destruction of Dresden, since all I would have to do would be to report what I had seen.

What I saw I could not remember.

I did not pick the book up again for a year.

———

This is how diagnosis occurs. At John Muir Medical Center, while nurses cared for me, the doctors inventoried all my symptoms and tried to figure out the root cause of my stroke. We needed to know whether or not I might have another one. And if so, why.

They first came up with a differential diagnosis process, which is a fancy way of saying "picking one diagnosis by process of elimination." The doctors listed all possible diagnoses that shared my symptoms, before narrowing down the list to root cause. A differential diagnosis procedure is like looking inside a refrigerator, seeing mustard, chicken, pasta, lemons, and milk, and deciding what meal you'll make for dinner that night. You could make mustard chicken on a bed of creamy pasta, or a cream of chicken soup, or maybe a chicken casserole—a list of possible meals all created from the same set of ingredients.

My symptoms included diplopia (double vision), headache, memory impairment, exhaustion. A cerebrovascular accident (CVA), another word for the death of brain cells due to lack of oxygen, which is another word for stroke, was a possibility. But since I was thirty-three, they did not put stroke at the top of the initial list. They added brain tumor and atypical migraine to the list. Vasculitis, too. Cancer was also a possibility—but because cancer cannot kill quickly on the spot, the doctors decided to screen for it later. They put that on a secondary list of possibilities. The more pressing alternatives—vasculitis or stroke—had to be confirmed sooner rather than later.

Other possibilities included a clot originating on the left side of the heart and then traveling straight into the brain. Or a paradoxical thrombus—a clot traveling from the extremities across the heart's central wall via a patent foramen ovale, or hole in the heart, into the arteries and then the brain.

In the beginning, you will recall, vasculitis was the most likely diagnosis because of my young age and because vasculitis is common.

Vasculitis inflames and destroys blood vessels, or veins, resulting in aneurysms and restricted blood flow. Aneurysms are balloon-like bulges in the walls of blood vessels. If you have an aneurysm in the walls of your blood vessels and it becomes stressed, your arteries and veins could explode. People with vasculitis have numbness or coldness in the limbs, a low pulse, high blood pressure, headaches, and visual changes. All things I experienced.

Vasculitis is frequently a possible diagnosis on *House M.D.*—it was mentioned in forty-four episodes over the lifespan of the series. Because vasculitis manifests in so many different ways in different parts of the body, with symptoms such as fever, weight loss, bloody nose, hypertension, and abdominal pain, it is a good first diagnostic guess. Vasculitis is a lot like an onion, to return to cooking; used in many dishes, it can manifest in different ways—fried or sautéed or pureed; sharp or sweet or caramelized; crunchy or crispy or tender.

A doctor relies on both experiential and learned knowledge in the differential diagnosis procedure. She tries to remember—with the imperfect and incomplete practical

memory recall of all humans, while navigating her own bi-
ases and errors—prior cases with similar clinical problems,
just as one would navigate one's own cooking repertoire and
know-how. This is why the road to an M.D. involves thousands
of hours of in-hospital work and exposure to all kinds of cases.
There is nothing that can replace experiential knowledge in
diagnosis.

As a result, disorders that are diagnosed most frequently,
that are most common, are assigned a higher probability. This
makes sense, as their symptom patterns are easily recognized
and the odds of a correct diagnosis are so high that no further
testing is required. For instance, if one has a runny nose and
a sore throat, it is more likely the cold or flu than it is scarlet
fever. If one's pantry contains peanut butter and jelly, then the
likeliest meal would be a peanut butter and jelly sandwich.
You will probably not make a peanut butter and jelly soufflé
or peanut butter and jelly baklava or peanut butter and jelly
muffins—even if you technically could.

But occasionally an unusual medical case comes along.
Like when someone who is only thirty-three years old comes
to the hospital complaining about sudden short-term memory
problems.

In cases where symptoms are more unusual, the process
involves further investigation and questions and theories
and tests. Doctors ordered a CT scan for me. That evening
in the emergency room the results came back, reporting
a lucency, a bright spot signifying dead tissue, in my left
anterior thalamus. There. There was the location of the
problem—not a migraine, likely not an abscess, and likely
not a brain tumor.

I was admitted immediately, and scheduled for an MRI the next morning to follow up on the CT findings.

Dr. Volpi came into my room and administered a short-term memory test. He told me to remember the words *apple, table,* and *penny,* then had a brief chat with me. We made small talk. I do not remember what we talked about. But I imagine it went something like this:

> *"So, Christine, how are you feeling today?"*
>
> *"Fine." (Apple table penny.) "How are you?"*
>
> *"Good. So, have you had any headaches or vision impairment today?"*
>
> *"I'm okay." (Apple penny.) "A little tired." (Penny.)*
>
> *"Your husband says you're sleeping a lot. About twenty hours a day? That's normal after a stroke. Sleep when you can."*
>
> *"Okay. That's an easy order!" (Apple?)*
>
> *"We're going to run some more tests to figure out what caused your stroke."*
>
> *"Okay."*
>
> *"For now, we're putting you on heparin to keep clots from forming. And you're on telemetry so we can track your vitals."*
>
> *"Okay."*
>
> *"Now, can you remember the three words I said at the start of our conversation?"*
>
> *"Three words?"*
>
> *"Yes. If you can't remember three, any of them is fine."*
>
> *"I can't—I can't. There was an ahhhh sound in one of them. I can't remember, though."*
>
> *"All right. That's okay—it's to be expected. The first word was apple. Do you remember the others? No? I'll be coming*

around again to check in on rounds. Is that your notebook?" Dr.
Volpi nodded at the Moleskine in my hands.

I didn't want to forget anything, ever. I was writing down
whatever I could in my diary. I wrapped my fingers around it.
Hugged it close. I nodded.

"Okay. I want you to start writing everything that happens
in that notebook. Write the time of day, too. That's going to be
your memory for a while."

I could do that, I told him. I could do that. And I did.

10 A.M. January 4, 2007:
Today MRA at 1 P.M.
Yesterday: MRI
Day before yesterday: CT scan, says something had
happened. And the doctor admitted me to the hospital. They
thought it was not a stroke but something else at the time.
- Angioscan
- In between: a bunch of therapists I forget.
- Ohhh I remember: a neuropathologist? Psychiatrist?
Yesterday. She says ought to have regular visits.

My journal would act as my short-term memory bank, it
turned out, for a long while to come.

In his notes following the MRI results, I saw that my neurologist had written, "Impression: subacute anterior left thalamic stroke in a young adult, essentially healthy. Vasculitis
and dissection need to be ruled out, also a new neoplasm."

A neoplasm meaning new growth, implying cancer.

Once I'd been diagnosed with stroke, the next steps were

to find out what caused the stroke. Was it my blood pressure? Was it totally by chance? Was it a PFO (patent foramen ovale)? My young age made it a priority to find the root cause. I had the rest of my life to live, one that without diagnosis would allow ample opportunity for recurring strokes.

The clot was in the left chamber.

It swirled, stalled for a split second, then entered the ventricle of my heart through the bicuspid valve, swam, gushed out the aorta into the arterial jet stream, through the left subclavian artery, then the left vertebral artery and basilar artery, and finally the posterior cerebral artery—like a car taking a series of off-ramps, each one closer to the city center—until the clot was in the recesses of the brain.

The tiny clot, which is also called an embolus or a thrombus, then made its way through the brain, along smaller and smaller arteries—picture boulevards and avenues, then residential streets, and then alleyways that no one outside a neighborhood could know—setting off static and images in its dusty path. It floated along until there was nowhere farther for it to travel. Stuck in the hub of the brain, it choked off oxygen from reaching the left thalamus. It was short-circuiting memory. It was splicing memory. It was killing the brain. The clot was stuck in time.

Dresden was in ruins.

Like Dory the Fish in *Finding Nemo*, I remembered little. But unlike Dory, I wrote all my thoughts and experiences—everything that happened—in my Moleskine. I detailed doctor visits to my room—what was said, and at what time they occurred, the activities undertaken.

I would have forgotten all of this otherwise. The nurses called me Forty-Seven, the last two digits of my room number in the DCU. I only know this detail now from reading my journal.

From my journals, I know that an assigned hospitalist, a neuropsychiatrist, a neurologist, and an occupational therapist visited me on a daily basis. This is how I know:

2 P.M., January 4, 2007:

Occupational Therapist

- we talked about going to therapy

- we walked around the hospital to the gift shop, and bought things, like a teddy bear, birthday card, and People magazine. Then tried to remember what I'd bought, later.

TODAY: echocardiogram of my heart, results normal

Then I was on some sort of steroid

Waiting on results of MRA

- am trying to get ahold of Mom and Dad

- am so freaked out and discouraged by the gaps in my memory—they seem so tremendous.

- at the moment, I'm excited about getting to shower!

8 P.M.

Mom and Dad called back. Tried to calm them down.

Now I am on heparin, with an IV. It's a blood thinner.

These notes to myself were the way I navigated my daily life. A reference guide. The shortcomings of my brain were bolstered by my journal.

There are things I have never remembered. Things I didn't write down. Gone. And even when I wrote things down, I still do not remember. I do not remember the teddy bear, birthday card, and *People* magazine at all, years later. Do I still have the teddy bear? Whose birthday? And did the therapist pay for those items? Or did I? Did anyone? Did we give them back? It could very well have been another person who wrote those notes down, save for the fact that it was in my own handwriting.

I don't remember having called my mother and father while in the hospital. I thought it had taken me weeks to get ahold of them. But my notes tell me otherwise—that I had indeed called them immediately after my diagnosis. That I talked to them the day after going to the hospital.

And yet there are things that I did and do remember. The name of my neurologist, Dr. Brad Volpi. And the toilet in my room. It swung out from beneath the sink. And then after using it, I would close the cabinet door and the toilet would then swing back into its dark space. The toilet delighted me, and because I was delighted, I remember the toilet.

How odd, memory. How useless, memory. How bendable. How breakable. How invented. How lost.

Things can be forgotten and life will still go on. What is remembered may not be critical. What is forgotten may be of significance. Time is dependent on memory. Time is unreliable.

But my memory book would change me, forever. Writing

was my way toward to my new self. My deficit forced me to write every single day; it forced me to construct legible words and sentences. And I can't help but think that describing my life also began to stir my imagination from deep within. Somewhere inside, things were happening—not just in my brain but in my mind, in my heart.

Later on, in recovery, I wrote down my feelings and thoughts in addition to the facts. My diary became, like a differential diagnosis procedure, a part of my eventual writing process—all the documented facts became a story. I pieced it all together into narrative. I had to recover to do so. I depended on my own experience and knowledge to cobble together themes and lessons and scenes and episodes into an essay, into this book. And in this way I truly became a writer.

———

I remained hospitalized over the next week as tests and monitoring to determine the cause of my stroke continued.

Apple, table, penny.

Adam repeated those words to me over and over. Tried to teach me mnemonic devices. ATP, he said. Think of something that will remind you of ATP.

Adenosine triphosphate!

ATP. Apple table penny!

I thought of another mnemonic device. "Apples on the table look pretty as a penny!"

Dr. Volpi came into my room again later, on his rounds.

"Hi, Dr. Volpi! Apples on the table look pretty as a penny!"

He cocked his head.

"It's my mnemonic device!"

"Well. Okay. That's a good method."

"I'm going to try to remember, Dr. Volpi."

We discussed tests to come, tests that had been done.

"All right, Christine. What are the words?"

I did not remember. I would not remember for months. And only then because I'd written them down. And because Adam remembered.

Apple.

Table.

Penny.

I used to have an excellent memory. The kind of memory where I could recall the page number of a book where my favorite quote was written. When I wrote English papers in college and needed to look up a line, I'd close my eyes and envision the line, which would then produce a picture of that page in my brain, which would then show me the page number. When I was bored in traffic, I memorized license plates of the cars around me. I memorized all my credit card numbers.

I memorized the credit card numbers of men who bought me drinks at bars. What would you like to drink? they would ask. I'm not looking for a date, I said. I don't care, they said. I'd just like to buy you a drink. They handed their card over. It sat at the edge of the bar. Their names were there. The names they had given me. I had given them a fake name. I'd see the sixteen numbers above that. *1234* . They were talking about their job in insurance or consulting or film production or technology. *5678*. They told me about their trip to Beijing and greeted me in Chinese. They were inviting me up to their room to smoke marijuana. *9101*. I was shaking my head and giving them the faintest of smiles, so as not to raise their ire. *1213*. I was looking at the numbers. *1234 5678 9101 1213*.

Would you like to come up to my room?

1234 5678 9101 1213.

No, I said, thinking, but I have your credit card number.

Oh, come on. It's good weed.

I took a picture of his face in my mind. His name. His credit card number. I shook my head.

I remembered people's names and never forgot them.

Instead of understanding chemical reactions in organic chemistry, I memorized the images in my head, which made it troublesome on exams when I was presented with a different kind of reaction to complete.

Furthermore, I took great pride in my superior memory.

And suddenly it was gone.

The erasure of an ability on which I relied heavily, which I considered a core part of my identity and intelligence, was shattering—for the next two years, when I said, "I don't feel like myself," what I meant was that I missed my memory, needed my memory, clamored for my memory, grieved my memory.

But in those first few weeks I was lost without knowing I was lost. I was searching with a deep belief that all would be well, not out of resilience or hope but out of ignorant bliss. I was in a hospital room, sheltered from the world, where nurses and doctors protected me from overstimulation, where everything happened on schedule, and where the blank white wall was not an acre of boredom but of great comfort. My world was that room, and in that room my struggles had little measured impact.

In that room my life inched forward in fifteen-minute increments. There was no boredom. Fifteen minutes was not enough time to become bored. It was only enough time to observe and feel and then forget. The hospital sheets were well worn and soft. The tug of the IV needle in my arm leashed me to the pole that dripped heparin. The blood pressure cuff tightened until it released, a sweet and regular choke. Food appeared on its own, whether on a hospital tray or from Adam's meal outings. The sheets changed without prompting. A cleaning lady, heeding her directive to be unobtrusive and efficient, sanitized the room without hubbub. The telemetry machine beeped as it monitored my well-being. A doctor stopped by without being called, on schedule. I took down notes. I slept. I woke up. I slept again. Light turned into dark turned into fluorescent lights turned into dark turned into light.

There was no future. There was no past.

Forgetting made many things easier. There was nothing to forgive, no grudges to carry. There was no anxiety and worry. There was no fear. There was chewing. There was sleeping. There was talking without remembering, without depth, words swirling around transactional items—what, when, where. There was carrying on.

The mind and brain are different entities. The mind, or soul, is abstract. The brain, flesh and neurons. But the functions of the mind and brain are linked—a marriage of partners, each one distinct but also related and connected to the other. Without the mind, the brain is an organ that has no way to express higher-order thinking. Without the brain, the mind starts to make up stories.

So it was that my mind gave my dead brain a narrative—it tried to explain what was going on, tried to make up the gap between injury and understanding. My mind got depressed, my mind struggled to make sense of things, my mind gleaned the lessons to be learned, and my mind gave me both pep talks and self-criticism.

My mind is what told me my brain was disabled—by depressing me, by slowing me down. By telling me I was tired, when my brain was overworked and needed rest. My mind was what held out hope and laid out a plan to come back. The mind told me I would be okay, because the brain needed to be well and to focus on healing. My mind told me to proceed hour by hour, day by day.

My mind is what writes this memoir now.

Now that my brain works again, my mind continues to try to understand. That the brain is an organ that died then came back online, albeit in a new way. That lives restart again and again and again. That this is a metaphor for all the deaths we experience inside us throughout our lifetime. That relationships end but life carries on. That tragedies occur but everything does not come to a stop because of a stroke or death or divorce or betrayal.

That even when I forget what happens, it doesn't matter—the world moves forward.

Writing saved my life.

I did not write that down. But of that I am certain.

4

I would be asked, "What does the thalamus do?" throughout recovery. Because we are asked to be experts of our own bodies when we become ill. And we do become experts, from firsthand experience.

I became familiar with my shortcomings as I recovered beyond the first weeks and understood what it was I could no longer do. And from that, I surmised the functions of the thalamus, one of the least understood parts of the human brain. Oliver Sacks, in his book *The Mind's Eye*, wrote, "One has to lose the use of an eye for a substantial period to find how life is altered in its absence."

I had not lost the use of an eye, but I did lose part of my brain.

This is what I knew after a few months of recovery:

I could no longer retrieve memories, even ones from long ago. I could not transform short-term memories into long-term memories. No matter how hard I tried, I could not remember apple, table, or penny. I could not remember things that had happened fifteen minutes prior. I no longer had witty comebacks. I would cry one minute and become angry the next. I, who before the stroke never cried in public, now wept when people said things that hurt me, much like a toddler. I could not walk down stairs without thinking left foot right foot left foot right foot, step step step.

To know this and not be able to change it was maddening. I felt helpless every single minute, and against this helplessness, I raged. It fueled my outbursts, it fueled my sadness, my grief; it made me punch the walls until finally, in exhaustion, I slept.

Even today, eight years later, I find myself staring intently at the stairwell as I place alternating feet on the steps. I have fallen down the stairs of my house more than once, more than twice, more than three times, since the stroke. I have a torn rotator cuff to show for it.

Without my brain, and specifically without my left thalamus, I had to figure out new ways to do the old things.

What I could not do, I had to ask for help to do. And I did so without practiced grace, shrieking like a toddler for what I needed, without the word *please,* and without specifying what it was—only saying no, not that, not that, and not that.

Instead of steeling myself when I felt sad like I used to before the stroke, I had to allow myself to cry. I allowed myself to feel sad. And this changed the landscape of my emotions. Over the years I became a less angry person, because my anger had actually been a cover for sadness and vulnerability. But throughout recovery, I was a very sad, very angry person. I screamed at the pharmacist. I screamed at other drivers. I had always had a short fuse, but that fuse got even shorter in the first year of recovery. I wanted everything, but nothing was good enough. Nothing was fast enough. Nothing could console me. What I wanted one minute, I did not want in the next.

The stroke changed my brain, which changed my mind, which then changed me.

What I could not glean from my well of experience, I had to intuit in the two years of recovery, especially in the first months, when it came to making decisions. I had always had a habit of second-guessing myself and double-checking my thoughts and theories against facts. In the wake of the stroke, I wasn't able to second-guess myself. I had to rely solely on my gut, just as I had while driving home from the grocery store two days after my stroke. When I could not re-member an acquaintance's name, I learned that the more I tried to remember, the more impossible it felt to recall. But if I relaxed and stopped trying to remember and let my brain come to a halt, somewhere in my gut there would come a name that felt completely random yet so persistent that I had to trust it to be true.

"Barbara?" I would ask.

And she would say, "Yes! I'm Barbara."

I would feel relieved. As if I'd driven along a darkened road without a map and somehow found my destination lit up in the night.

Intuition drove me forward.

Intuition propels us to do something without reasoning, without knowing how or why. It sees patterns and possibilities in the information received. It is what's meant when people "read between the lines." It is trusting impressions, symbols, and metaphors more than what is actually experienced. To know the way home without a map, because it is home and no map is needed. Intuition is, as Steve Jobs once said, "more powerful than intellect." It is processing feelings and thoughts in what I feel is a primitive way—leaping to a conclusion without going through steps, based on experience and instinct and deep wisdom. It is meeting someone and liking that person immediately and knowing that you will become friends and then you become friends.

———

In the interim, I lay in my hospital bed, exhausted and nearly mute, much like a newborn sleeping on and off. Every fifteen minutes the blood pressure cuff checks my blood pressure. The IV drips anticoagulant and fluids. Adam brings me meals. Adam fields my phone calls. I send out e-mail messages from my BlackBerry, ones I will not remember having sent. Nurses bring medication. Doctors monitor my progress and strategize on diagnostic tests. I am whisked around the hospital either in a wheelchair or flat on my back in my bed, to MRIs and ultrasounds. I do not move a muscle unless I have to. I do not know what time it is. And I am not bothered by this lack of awareness.

I make no decisions. I cannot make a single decision, for months to come.

The thalamus is a primitive part of the brain, its functions not as well documented as other parts of the brain, such as the prefrontal cortex. Each side of the thalamus is the size and shape of an unshelled walnut—about three centimeters in length, two and a half centimeters across, and two centimeters high; the left and right thalami combined are about the same size, weight, and shape of a marine iguana's brain.

This neural structure is located at the top of the brain stem between the cerebral cortex and the midbrain, with extensive nerve connections to both. It is in the center of the brain and acts as the hub of the brain for information flow—a traffic circle, so to speak, or an Internet router. As such, the thalamus has multiple functions. It regulates consciousness and sleep when it communicates with the cortex. It is involved with sensory and motor relay. And it helps the mind make decisions.

The thalamus receives a signal whenever we see, touch, taste, or hear something, then directs it to the part of the brain that can process the information, whether to the prefrontal cortex for strategic decisions or to the primary motor cortex for movement. It helps the brain make sense of the information coming in. That the bright spot in the sky is the sun and the sun is a star and it shines during the day and today is going to be hot and maybe leave the jacket at home. That this slippery fabric is satin like the wedding dress worn years and years ago and then never worn again and maybe I should never spend such money on something I will wear only once. That the sweet and crispy thing called a pastry that I bit into and which cracked into a thousand layers in my mouth and then burst into sweet and salty and buttery is known as a *kouign amman*, and I have loved it from the day I first tasted it at a farmer's market on a cold morning in November as the light peeked in and out of the sterling clouds rolling overhead and forever will I crave that pastry on a cold day. And that that sound is a song in the voice of Annie Lennox and on certain gray mornings I played her cover of "A Whiter Shade of Pale" (the room was humming harder) while staring at the ceiling wondering how I would get up that day (as the ceiling flew away) and go to work and drive sixty miles (a whiter shade) each way (of pale), sixty miles of not driving (she smiled at me so sadly) into the center divide, and then I finally did get up. I got up. My body moved.

The thalamus plays a role in controlling the motor systems of the brain responsible for voluntary body movement and coordination, like when I see a red pot and I realize that that is the pot I need to boil my pasta, and then I decide to reach out to the pot—the thalamus coordinates and relays my mind's desire and the decision made within my cerebral cortex to the motor cortex, which then communicates with my body's muscles, which then results in my arm swinging out to the pot, my fingers grasping its handle and pulling the pot up and off the shelf.

Each part of the thalamus relays messages from one part of the brain to another. It takes outside information and figures out which part of the brain ought to receive this data, thus producing conscious perception; it is often referred to as the gateway to the cortex.

It is why I saw shapes and colors but did not know what they were. It is why I saw food but did not know what kind of food, or the name of the food. It is why I could taste something but have no appetite for it—no triggers to tell me I'd liked Fuji apples for years, hated bell peppers, no triggers to tell me ice cream made me happy and think of days my mother rewarded me with ice cream for a job well done at school.

My medical files note a "4mm x 9mm area of acute left thalamic ischemic infarction." About a dime-sized dark spot of dead tissue, objectively a tiny piece of tissue. In any other part of the brain, such as the frontal lobe, a small spot of dead tissue would have meant a minor stroke, one that I might not have noticed having. But that tiny dead spot changed my life; such a scar in the thalamus is large in context.

To return to the thalamus-as-router metaphor: damage to one part of the thalamus cuts off communication to a significant part of the brain, and too much damage results in a permanent coma; in my case the stroke occurred in the dorsomedial thalamic nucleus, which communicates with the prefrontal cortex. And thus my prefrontal cortex went offline, as if it itself had been ravaged. I lost my ability to make decisions, organize, strategize, and solve problems.

In this way, I lost my personality.

The prefrontal cortex is a "newer" part of the brain—it synchronizes thoughts and actions with internal goals and is thought to be where personality resides. It is where executive function takes place—where our brain differentiates between better and best, good and bad, same and different, where it sees future consequences of present actions, where social control operates. The PFC is where short-term memory is stored.

The prefrontal cortex is like an organizer. It has the unique ability to drift offline, going between past and future, to time travel as it coordinates actions and goals with concepts like morality. To look at a Thanksgiving turkey and remember all the Thanksgivings in my life. To then recall the first turkey my mom roasted, slathered in mayonnaise, per the suggestion of one of her friends. How my mother wanted to give me what I wanted, even though she had no idea how to cook a turkey because she had never grown up with Thanksgiving. How I found it dry and unappealing. How we then had roasted chicken the next year. And then my mother's amazing North Korean chicken soup for the years following, because we loved that soup, and chicken is poultry, so why not. How Adam loved Thanksgiving and roasted thirty-pound turkeys for the two of us. How the carcass made amazing stock. How I eat the sides and not turkey on Thanksgiving, because turkey never tastes great. To the turkeys my friend raised on her urban farm. How she defeathered them. To feathers, to peacocks that wandered my childhood town in Southern California. To goose down and duvets and warmth.

The ability to delay gratification resides here, too. And yet, luckily for me, the newer parts of the brain rebound better from damage. Why has not yet been determined.

But the abilities of my prefrontal cortex did come back.

But before I recovered, I slept. And slept. And slept. I slept twenty hours a day. In the first weeks and months following the stroke, I continued to sleep, regardless of place and time and setting. On the couch. In my bed. In the car. I was exhausted, because my brain was exhausted. I slept dreamless then, and dreamless for months. Or maybe I dreamt, but I do not remember having done so. I slept to recover.

———

Without my thalamus, my brain went offline. It retained nothing. Thoughts slipped in and out of my head as if I were trying to open jars with my hands greased in oil.

In those early days, conversations were circuitous.

"Hi, how are you?"

"Good, Christine."

"How have you been?"

"Christine—"

"Yes?"

"We had this conversation fifteen minutes ago. I'm fine."

"Oh, but how are you? I want to know."

"Great. Just got off work to visit you at the hospital."

"Oh! Thanks for coming! Would you like a cookie? Someone brought some in. I don't remember who. But there are cookies here."

"I already had one."

"Really?"

"You offered me one earlier. I was the one who brought them."

If I had been back to normal, I would have then started to fret about the future and to plan, to worry about whether or not my absent-mindedness had offended, would affect our relationship, made the other person feel less significant. I would mitigate any shortcomings—apologize, make amends, compensate, or crack a joke to diffuse tension. But I was not normal. I was unable to consider the future. I was surprised. And then I was not. And then my mind went blank. I did not even wonder what it was I should have said in response.

I grew quiet at such moments, at such forks in the interactive road. If there were other people in the room, in my silence they would begin to chat to cover with a sea of conversation that void.

At one point during my hospital stay, my friends from work, from the software company Adam had started, visited me. Just two weeks prior, I had been a touchy, take-no-prisoners colleague.

There I was, in the bed. And they greeted me by saying, "You look completely normal!"

There was, I suspect, relief in their voices. That they didn't have to look at someone whose face resembled melted wax. That I didn't talk like someone whose mouth was full of marbles—that they didn't have to translate.

We made small talk, like any other small talk. Thank you for the flowers how are you wow you look great work's good how are you how are you how are you how are you?

And then my neuropsychiatrist stepped in, mid-visit. The room went quiet.

"Hello."

"Hi."

"Do you know who I am?"

"No, but it's nice to meet you."

"Open up your journal."

I opened up my Moleskine.

"What time is it now? What is the time of your last entry?"

"It's ten thirty-five. Oh wow! I met you twenty minutes ago! Oh! You're my neuropsychiatrist. That's who you are."

"Fuuuuck," said my friends, breaking their silence. "Wow."

Later, one of those visiting friends said, "I thought you were fine until that moment. Until then we didn't realize you couldn't remember anything."

And really, at that early point in recovery, neither did I.

I didn't know what I didn't know. What I couldn't do.

5

Short-term memory dominates all tasks—in cooking, for instance: I put the water to boil in a pot on the stove and remember that the water will boil while I chop the onions. I will put the sauté pan on the stove to heat up the oil for the onions, and I will then put the onions, which I will remember I have chopped, into the oil, which I remember I have heated for the onions. I will then add tomatoes. While the onions and tomatoes cook, I will put pasta in the water, which I remember I have boiled. I will know that in ten minutes I will put the cooked pasta into the tomato and onion stir, and thus have a simple tomato pasta meal.

If short-term memory is damaged as mine was, it works more like this: I put the water on to boil. I heat up the oil in the sauté pan. I chop the onions and then wonder for what it was that I chopped the onions. What might it be? I wash my hands, because I might as well—my hands are covered in onion juice, and my eyes are tearing. I return to the stove, where the oil is now scorching hot. I wonder what on earth it was I was cooking, why the sauté pan was left this way. I turn off the heat under the oil. I sigh and go upstairs. I forget everything I just did like a trail of dust in wind. Two hours later, after a nap, I return to the kitchen to a pile of chopped onions on the chopping block. The pan is cool but scorched. And I again wonder why. But mostly, my eyes turn to an empty stockpot on the stove, the burner turned on high. There is nothing in the stockpot, not even water. This happened over and over again in the months following my stroke. So I stopped cooking for a year.

Short-term memory is like an administrative assistant for the brain, keeping information on hand and organizing tasks—it will figuratively jot down a number, a name, an address, your appointments, or anything else for as long as you need to complete your transaction. It stores information on a temporary basis, on Post-it notes, before deciding whether or not to discard the memory/Post-it or move it into a file cabinet for long-term memory storage. Everything in long-term memory finds its way there through short-term memory, from the PIN for your ATM card to the words to the "Happy Birthday" song to the weather on your wedding day. In fact, you are exercising short-term memory now, by keeping track of what you read at the beginning of this sentence so that you can make sense of it at the end.

When short-term memory is damaged, it cannot track sentences. It must read the paragraph over and over again, because by the end of the sentence or paragraph, it will not remember the beginning. And because it does not remember the beginning, it cannot make meaning out of the entirety.

I look at a restaurant menu. I read each item, and when I get to the end of the list, I cannot remember what was at the beginning. I reread the menu. I get to the bottom of it. My brain gets tired, short-circuits, and all I see is random words. I cannot connect my appetite to the words. I cannot remember what food tastes like. I cannot connect the ingredients, "hand-cut green noodles with chanterelle mushroom ragù and gremolata," into a whole. I cannot put together noodles and mushrooms and chopped herbs in my brain. I cannot connect those flavors into a picture, and I cannot connect them to my appetite, because I have no memory. I only know I am hungry, because I am light-headed and listless.

I put down the menu. I ask for a hamburger if I am dining alone. I ask my companion to order for me, if I am not dining alone. I always request hamburgers, because nearly every restaurant offers hamburgers, and because I cannot parse a menu and hold all the possibilities in my head in order to make a decision.

I am surprised, every time, when the hamburger arrives at the table, because I do not remember having ordered it. I chew it mechanically. There are no images flashing through my head reminding me of the first time I ate a hamburger, or all the barbecues I've attended, or the time after marching in the Rose Parade that I ate Burger King because Burger King gave out free burgers to participants at the end of the route. No. There is just blank space. There is chewing. Swallowing. The end of hunger.

When short-term memory is damaged, it will not retain new names. I do not remember someone who popped her head into my hospital room a few minutes ago. I do not remember the receptionist in the doctor's waiting room. I do not remember who visited me in the hospital the day prior. I do not remember who gave me the flowers in my room. I have to write all these things down in my notebook, so I can refer back to it later.

If short-term memory is damaged, it may not be able to move things into long-term memory, because it takes time, even if not much. It can take about a minute for the memory to be retained. But with age or injury, our brains have less time to successfully move new information to long-term memory. As a result, it is difficult to recall the details of recent events. I see a book at the bookstore, and I buy it because it looks interesting. I go home and see two copies of that book on my bookshelf because I have bought that book over and over again.

I do not remember so many things that happened. I do not remember who was in my workshop the semester I returned to school before I was fully healed, returning because all I wanted to do was finish my degree. I do not remember the woman who befriended me in the wake of my stroke, who then months later wrote me a breakup card because supporting me, she said, was "too much." I find the breakup card years later and look at the date in befuddlement. I do not remember printing my MFA thesis onto special paper and then assembling it and turning it in. I do not remember the names of any doctors at the hospital. I do not remember room numbers. In addition to not cooking, I do not even go grocery shopping for a whole year, because I forget what it is I have to buy, and if I write a list down, I forget where I put the list.

In the wake of my stroke, I remembered the names of people I'd known for years, even if I couldn't remember the names of doctors I had just met. I recognized my best friend, Mr. Paddington, and my husband, Adam, and all my girlfriends, and greeted them. But when they'd leave the room and return, I would greet them once again, as if they hadn't been in that same room just fifteen minutes prior. I knew who they were, but I had lost track of time. My short-term memory was unable to move things into long-term storage.

Long-term memory, also known as reference memory, is remembering anything that happened more than a few minutes ago. Whereas short-term memory lasts well under a minute, information in long-term memory can last indefinitely, from a few days to decades to an entire lifetime. It is remembering my first trip to Disneyland, and how the night before, my father, brother, and I laid out a strategy—that we would run to It's a Small World first, because it was at the back of the park and the lines would be short at opening. It is remembering my first day of school in New York City—how I was born in the United States but didn't yet know English, because my parents didn't want me to have an accent like they did—their diction made them suffer, and they did not want me to be similarly marked. How in 1977 the teachers had no experience with a non-English-speaking child. How I cried my first day of school. And how they locked me in the bathroom stall until I would stop crying. And how I did not. And how I was locked up the second day and then the third. And as I moved through grade school, how I was the only one of two Asian

American children in the entire school. How when a new immigrant from Korea showed up, my beloved teacher turned to me and asked me to teach him English. How by then, at age seven, I'd forgotten Korean. How I picked up a Crayola crayon, pointed to it, and said, "Yellow." How I wondered why the teacher could not think of doing the same.

How I would command a sofa and call it a boat—"Get up here, Richard!" How my brother would say, "I have my own boat," as he got up on the other sofa. How I replied, "That's not a boat! That's a sofa. Get on my boat or you'll drown!" How he said, "But this is a boat." How I stared him down and said, "That is a sofa. You are captain of a sofa. I'm captain of a boat. Get on my boat if you want to live!" And how he did. And how I then told him, "Now, row. I'm the captain. You're the rower." How he and I were the best of friends. How one day he grew up and made other friends.

Long-term memory allows me to remember the day I got my driver's license. My high school graduation in 1991 and how I wore black opaque hose and white pumps. Because that was the style back then. Really. It allows me to remember the organic chemistry exam I failed. How on the morning of 9/11, my boss called and woke me up screaming, "We're being attacked! Stay home. Buy water! Don't come into the office," before hanging up, before I could even respond. All the times someone told me to "go back to your country," and all the times I screamed back, "I was born here! This is my country!" The first time a short story I wrote got published. How I was alone in the house, and I opened the envelope, and held my breath, and I knew then that I should keep writing.

There are two kinds of long-term memory: explicit and implicit.

Explicit memory, also called declarative memory, is the conscious recall of previous experiences and information. It's what we most associate with things we call "memories." It's remembering getting my driver's license, and also the driving lessons and my nerves and anxiety. It is remembering my score on the driving test.

Explicit memory is the encyclopedia of our experience. It is having been born in New York City. It is knowing that my mother was a nurse and my father an engineer before he was a small business owner, first of a gas station and then a barbecue restaurant and then a water store. It is knowing that they arrived in the United States in 1969 after the passage of the 1965 Immigration and Naturalization Act, which allowed immigration to the United States from Asia for the first time in forty-one years. That they were recruited to the United States under category 3 of that act, which favored scientists, engineers, nurses, and medical doctors.

It is remembering the name of the New York City preschool I attended. What I wore on my first day: a red dress with white polka dots and black Mary Janes. It is remembering that bathroom stall, and how I kicked the door all day long with my patent leather black Mary Janes. And the next day. And the day after that.

There are two different types of explicit memory.

Semantic memory is composed of the knowledge we have accumulated throughout our lives. It is the textbook of our lives. It's knowing that the San Gabriel Mountains surround the valley. That the highest peak in that mountain chain is Mount Baldy. That rock cod reside in seaweed, at lower depths than mackerel. That Richard is my little brother. That my first preschool was located in Queens, New York.

Episodic memory is about the experiences and events of our lives. It helps us time travel. Episodic memory is about the Sundays my family spent hiking the San Gabriel Mountains, how I kept gasping for air, and how no matter how often we went, I never found it easier. How I went fishing with my father for rock cod as a child. How we had to cast our lines differently than with other fish—a long line that we let go on the reel until we felt the weight hit the ocean floor. How I loved the rocking boat and the ocean and waiting for the tug on the fishing line.

Semantic memory tells me I was born in New York City. That New York City is the largest city in the United States. That I entered UC Berkeley as a double major in English and molecular cell biology. That I graduated as an English major with a minor in Asian American studies. That I earned my MFA at Mills College. That I spent twenty years working in tech. That I'd married Adam in San Francisco. That we went to Brazil for our honeymoon. That Brazil is in South America.

But episodic memory says that I grew up in New York City, eating pizza for lunch under the El in Queens. That I crumpled under my parents' wishes for me to become a doctor. That they thought a writing career was too daring. That I pursued it anyway. But that first I worked in tech to earn a living, because before I did become a writer, I too thought it was too daring. That during the hora at our wedding, my friends hoisted me with so much strength, my body left the chair, and that I laughed with both surprise and joy. That I barely clung to my end of the white cloth, the other end of which Adam grasped. That Adam and I went to Brazil for our honeymoon a few days later, and the first meal we had upon arrival was at a *churrascaria*. Where we scarfed down meat and filled our bellies in fifteen minutes, and laughed when we got the bill for all-you-can-eat beef and chicken and lamb: it was less than nine dollars for both our meals. Where we woke up to the sound of the ocean below us, which reminded me of all the times I went rock cod fishing with my father, liberating and strange, the waves crashing repeatedly.

In the wake of my stroke, my explicit memories were largely unavailable. I could recall modules. But I could not splice them together, even when prompted, which in the neuroscience world is called priming.

"I went to the store," I said.

"And?"

"I don't know. I went to the store."

"And you were trying to buy groceries?"

"Oh yes!" I said. "I couldn't buy any!"

"And?"

"I forget. What was the point of this story?"

"You tell us. What are you trying to tell us?"

"I don't know. I just went to the store. I guess that's it."

I spoke only in images.

"One time we went camping in the winter. We woke up in the morning, the tent covered in snow." I could remember no more. I had a picture, and that was it. I could not say what happened next. I could not remember the point of my story. In this way, I realized that memory and storytelling are modular, vignettes that I had to weave together with an overall theme and story. That I did go camping in the winter. That we did wake up in the morning to a snow-covered Alpine forest. That it was difficult to collect water from the lake. That we boiled the water for oatmeal. That it was beautiful. That the forest was silent, save for the crunching of the snow under our boots. That it was Adam's and my wedding anniversary that weekend. That Adam and I used to go backpacking. That I missed doing so. That the best times we had were when we were alone, away from the material world.

The stories were gone. Left were the images, the concrete details.

As a prose writer, I found it frustrating to not recall a story. To not retrieve memories in their entirety. To have only an image to hold. To have lost imagination, which is a way of lying, which is a memory exercise in and of itself.

———

Implicit memory is an unconscious, unintentional form of memory. Remembering my hikes as a child is an explicit memory, but improved mental endurance from the hikes, where I learned to take smaller steps and count in multiples of ten and tell myself that the pain would not be forever, is implicit memory, much like knowing how to ride a bicycle. Once you learn how to ride a bicycle, you need never recall each motion that has to be completed to rotate the petals, to maintain angular momentum so you can keep upright. Implicit memory is knowing how to type and not having to recall the placement of each key. Or knowing how to play a piano without having to remember which key to press. It is knowing how to wash one's face or body. It is knowing how to eat. That you must chew and swallow. It is knowing how to fish. Understanding what it feels like to be on a boat. To go with the boat's listing. To stand up on my toes without thinking. To do the same in a subway car. It feels like instinct.

It is coming to understand the critical role of communication from the experiences I have had, such as being locked up in that bathroom stall on my first days of preschool. That if I do not end up communicating my wants and fears and needs in words as opposed to screams and punches and kicks, I will end up in a place very similar to a bathroom stall. That words will unlock me from that stall. Only words will save my life. That rage will bring me attention, but language will bring me out from isolation.

All this would make me a loudmouthed little girl, who believed that her perfect, unaccented English would help her feel like she belonged, that it would counter her little Korean face, which was perfect but made her feel less than perfect throughout her life. And this loudmouthed girl would grow up to be a loudmouthed woman, who would in turn become a writer. Words will save my life.

Implicit memory, also known as nondeclarative memory,

is knowing how to drive home. To have done it so many times that even if my brain was damaged, I could find my way. Or how to drive at all, once you've learned to drive. It is taking all the driving lessons and the driving test, and then not being able to think about how to do it again, but doing it anyway.

It is being able to punch in the correct phone number for Adam, even though I couldn't recall the number if asked.

This nondeclarative memory drove me through recovery, though I did not know it at the time. All the lessons of my life to date came to light. This kind of memory retained all the things I knew how to do deep in my bones and helped me navigate a landscape in which I was otherwise lost.

I couldn't remember, I said, but the implicit memories were there. Those Sunday hikes I took as a child with my family taught me to manage that which was not easy, and that at the end of each path was a reward, even if it was solely the achievement of having completed a journey. The lessons learned and hardships overcome were resources invaluable to my recovery. As was the value of being creative in the face of adversity, like picking up a crayon and saying "yellow" and another and saying "purple" in order to teach another person English.

My childhood stories became lessons.

In this way my life had prepared me for such an emergency. It had taught me to anticipate hardship, to push through pain, to wage battle.

But my stroke would teach me things too, among them the value of taking a break from the unending pressure to be perfect.

My childhood would help me survive; in turn, surviving would erase my childhood.

When episodic memories are woven together and combined with semantic memory—there is a gradual transition when episodic memories lose their association to events over time and become semantic memory—they create autobiographical memory. Autobiographical memory is what I am using to write this memoir. Autobiographical memory is creative, consisting of vignettes that we remember from our life. The most vivid autobiographical memories are from emotional events— like my first memory, of throwing a snowball.

Heightened emotion can increase the likelihood of memory storage and retention—I remembered my neurologist's name when I could remember no other doctor. Maybe it was because he had the same surname as a childhood friend. Or maybe it was because he was kind. Emotional memory is how, in hindsight, I remember certain things and not others in stroke recovery without prompting. It is why certain dates hold significance in my mind. Why the sound of the 7 train on the tracks through Queens gives me such comfort.

It is why I try to make positive memories for my daughter. Emotional memory gives us the memories that haunt. I've always hated Halloween, because I remember all the arguments in my household; because my parents were immigrants, my costumes were never the caliber of other children's. How for some, Halloween is Samhain, when ghosts travel from other worlds to ours. How I feel so haunted on Halloween, my ever-vigilant self so ill at ease among ghosts and scary surprises, because of cultural superstition and the fact that my parents had one of their biggest fights on Halloween. I now try to celebrate a holiday of which I am not fond, so that my daughter can form her own associations with October 31—a decision I made based on my emotional memory and with the intention of creating happy emotional memories for my child.

The thing with memory is that it changes, it shifts, it is subjective. For me, memories are like reading a novel—you have a different emotional connection to *The Great Gatsby*, for instance, upon first reading than on subsequent reads. When in high school, I saw it as an illustration of the American dream. In college, I saw the bits and pieces of it, the car for the material dream of America, the language to be admired. As an adult, I related to it as a psychosociological snapshot of wealth, situated as I was in the Silicon Valley, surrounded by newfound wealth. And I saw the issues of race within.

When I read Junot Díaz's *Drown* for the first time, I was entertained. I thought Yunior was a cad. When I could read again, I read his novel *The Brief Wondrous Life of Oscar Wao*. What a guy, I thought—Yunior's got issues. And again, Oscar, the nerd boy marginalized on all sides, extinguished. And then later in life, after having experienced infidelity in my marriage, having been betrayed, I experienced rage and disdain toward Yunior, whose flaws included cheating on girlfriends. How could he? And then finally I felt compassion for him. The cheater is bereft of power. The cheater is trying to find power. The cheating is not the problem.

The writer is trying to find power. The writing is not the problem.

The girl is trying to find power. The problem is her mind, she thinks. The problem is that everyone else can exercise to exhaustion and she cannot. The problem is that she must not be strong enough in mind. The problem is her body. Her body must be ignored. Pain must be ignored. The mind must override everything. The mind has decided the body does not count. The mind has decided to punish the body. She will starve her body. She will purge her body. Her mind will count every calorie that is eaten.

The brain will excel in school. The brain will be the star.

The mind will make up stories for the body's deficits. The mind will say her body is a failure. The mind will favor the brain. The mind will fault the body for the stroke, of which the brain has become the victim.

The mind, without the brain, will finally have to learn to forgive the body.

6

Warfarin, the brand name of which is Coumadin, is the most widely used anticoagulant in the world.

In the 1920s, cattle on the prairies of North America began dying of internal bleeding with no obvious cause. Researchers found that the incidence of bleeding increased when the climate, and therefore the sweet clover hay on which the cows dined, was damp. There happened to be a series of wet summers in those years, exacerbating the circumstances. Eventually, a Canadian veterinary pathologist named Frank Schofield discovered the hay to be infected by the molds *Penicillium nigricans* and *Penicillium jensi.*

(Schofield also happened to be a key figure in Korea's independence movement from Japan, a movement that had a deep impact on my own family; he spent his early veterinary years teaching in Korea, becoming increasingly outspoken about Japanese colonialism and secretly photographing police and military and the aftermath of massacres. He treated Koreans punished by Japanese torture and offered his home as a sanctuary to such survivors. After retirement from work in Canada, he returned to Seoul to teach at my parents' alma mater, Seoul National University, from 1955 until his death in 1970. He was buried in the Patriot's Section of the Korean National Cemetery as the "34th Patriot." When I discovered this fact in my research, I was struck by how many impacts this one man—a stranger—had had on my life.)

Normally, sweet clover hay, with its succulent stems that make for easy mold contamination, would be discarded if it spoiled in storage. But the 1920s were not a normal time.

There was an agricultural depression in the 1920s, a period in which the rest of the United States enjoyed great affluence. During World War I, American farmers prospered—Europe was unable to produce enough food, so the United States' farm production expanded to fill the wartime gap. U.S. farmers took out mortgages and loans to buy more land to meet demand. But after the war, in the 1920s, Europe's agriculture rebounded, leaving the United States with a surplus and overextended farmers who could not shoulder the burden of their loans.

Given the financial hardship at the time, farmers could not afford supplemental feed; they gave their cows the spoiled hay. This then resulted in the hemorrhagic ailment called sweet clover disease. The situation only worsened with the onset of the Great Depression.

In 1933, ten years after the disease was diagnosed and the cause pinpointed to damp sweet clover hay, a Wisconsin farmer named Ed Carlson drove two hundred miles through a

February blizzard to his local agriculture investigation station in a desperate attempt to get help. He ended up at the office of Karl Paul Link, the only occupied office at the time. Carlson handed Link a pail of the uncoagulated blood from his dying cows. In his truck was one of his dead heifers along with one hundred pounds of old sweet clover hay.

Link's students Harold A. Campbell and, later, Mark A. Stahmann, experimented with the blood and the hay, produced after an especially wet 1932 summer. Although the cause of the disease had already been confirmed years prior, the actual active compound in the clover hay had yet to be discovered. In 1940, after years of intensive work, dicoumarol was isolated. Natural coumarin oxidized in moldy hay— and *Penicillium nigricans* and *Penicillium jensi* metabolized the coumarin into the substance that would be known as dicoumarol.

Dicoumarol is similar in structure to vitamin K, and when livestock consumed dicoumarol, it inhibited vitamin K production. And vitamin K, prevalent in food such as seaweed, kale, spinach, and chard, activates prothrombin, the very thing needed to produce clotting at the site of tissue damage. Dicoumarol prevents this process and provokes hemorrhaging.

The Wisconsin Alumni Research Foundation (WARF) funded Link's work, and patent rights for dicoumarol were given to WARF in 1941. The drug was named warfarin in recognition of WARF's funding.

Warfarin was first used as a rat poison in 1948 and is still used as such today. Rats have developed a resistance to it, however, and so its use is now declining.

In 1953, Stalin was supposedly poisoned and killed with warfarin, which is tasteless and colorless and water-soluble.

In 1954 it was approved for medical use in humans under the brand name Coumadin. When Dwight D. Eisenhower had a heart attack in 1955, he received Coumadin as part of his treatment. After that, warfarin was widely accepted as an anti-coagulant.

It was part of my treatment, too.

———

One week after admission, on January 6, 2007, I left the hospital and went home.

I sat in a wheelchair by the curb outside the automatic doors of John Muir Medical Center, waiting for Adam to drive up in the car. I gripped the blue personal-belongings bags filled with stroke literature, discharge instructions, and the cards and gifts friends had sent me in the hospital.

Calm and quiet as I must have appeared on that curb, I was losing it on the inside.

I'd been expelled from the womb of my hospital room with its constant temperature and filtered light and my every need met, and now I'd surfaced into a world where the sunlight and noise and movement, even the intermittent cool January breeze, assaulted me at every turn. Everything felt new and unfamiliar, and I recoiled.

The sun shone brighter than any fluorescent lamp, lighting up the remaining leaves on the trees as if they were shimmering shards of glass.

The breeze brushed my cheek with irregularity, each touch

an unwelcome surprise. My brain yelped with astonishment and shock.

The light wind made the leaves on the sidewalk scuttle, like frantic crabs on tissue paper. The sound buffeted my ears.

So much sensory input—an overload.

I wanted to scream, I wanted to cry, to laugh, to whimper, but mostly I wanted to run back inside the hospital, to those white walls, to that bed, to that muffled quiet of room 47.

It felt like I'd been gone from the world for months.

This, I realize in hindsight, must be a little like what it feels like to be an infant coming into the world.

But I could not run back inside. Adam was driving up with the car. And he ushered me into the passenger seat, holding my hand, excited to be taking me home.

I registered the click of the buckle. The pressure of the lap seat strap against my belly and the shoulder strap against my chest. The texture of the seats and the colors of the dashboard lights. The idling engine louder than any oxygen tank. The stereo, blasting noise, more audacious than an MRI.

As loud as the world felt while I waited on the curb, it became supersonic when Adam started the car and drove.

When the car moved forward, the trees and telephone poles and other cars accelerated, while the lights glinting off chrome and windshields multiplied. I stared at Mount Diablo, the stillest thing I could see, so that I could center myself in the whirlwind. Even then, it inched forward, and as we passed it, away.

The car stopped at lights. The traffic lights turned color, and each of the colors meant something important. The street signs had words on them. They were white on green, then white on brown, as we went from town to town. The lurching was immense, as Adam pushed forward, his race-car training intact. I wanted to stay still, but I moved, the world moved, and the world moved me.

The power lines danced.

It hurt. All of it hurt, like staring straight into the sun.

We got on the freeway, and everything went faster. The car ate the road before us. My brain felt like it was screaming.

I closed my eyes. I wanted to cover my ears and cry. I wanted to be swaddled and held tight. I wanted the air coming out of the vents to stop blowing. I wanted the errant hairs from my ponytail to stop moving and touching my face.

"Please turn off the music," I said. "Please turn it off."

Adam loved playing music while driving, and normally I did too, but right then I could not deal. I could not find an ounce of strength inside myself to allow him that pleasure, even though I wanted to reward him for taking care of me and for being so patient and kind. For all the lunches. For holding my hand. For sleeping on a four-foot-long bench each night in my hospital room. For driving us home.

But it was too loud. I could hear screaming in my head.

Adam clicked off the stereo. Without taking his eyes off the road, he said, "Okay. It's off. Honey, please stop screaming."

I wouldn't be able to listen to the radio in the car for another year.

The hospitalist discharged me with the following notes in my chart:

1. Left thalamic ischemic cerebrovascular accident presenting as memory deficit. Being discharged on Lovenox and Coumadin. Lovenox is to be continued until INR is more than 2.0.

2. Transesophageal echocardiogram showing trace mitral regurgitation and tricuspid regurgitation and interatrial septal aneurysm with PFO shunt, mild-to-moderate with Valsalva maneuver. Recommend outpatient followup with cardiology to discuss further options, i.e., anticoagulation versus closure.

3. History of migraines.

I was on a number of anticoagulant medications at the hospital and thereafter.

During my stay, I was put on a heparin IV. Heparin is an injectable anticoagulant, a blood thinner to prevent clotting. It is on the World Health Organization's list of essential medicines, the most important medication needed in a basic health system. Discovered by accident in 1916 by second-year medical student Jay McLean under the direction of the scientist William Howell at Johns Hopkins Medical School, heparin is one of the oldest drugs still in widespread use. It has been used since the 1930s, after it went into clinical trials, to prevent clotting.

In the hospital I wandered the halls sometimes by myself, but often with therapists, wheeling my heart monitor and heparin on an IV pole. The nurses allowed me to wear my own pajamas, so I was dressed in my black WHO'S YOUR BUNNY? T-shirt and pink gingham pajama pants. That IV pole feeding me medication that lowered the viscosity of my blood was the thing that grounded me, that made it very real that I was not all the way well.

On my solo jaunts I lost track of time and of my where-abouts within the hospital. It wasn't that I was restless. It was that I hadn't realized how brain-damaged I was when I stepped out of the DCU. I always intended to go only halfway down the hallway outside of the DCU, but once I got there, I forgot my intentions and kept going, because, well, I could. I once found myself in the labor and delivery wing of the hos-pital, outside the bounds of telemetry. The nurses lost track of my heartbeat. When I did return to the DCU, I was often scolded—please don't leave. They had been alarmed.

"I won't do it again," I said, but then I forgot and wandered out again.

Upon discharge, along with Lovenox, I was put on the blood thinner Coumadin (warfarin). Coumadin's major is-sue in widespread application is dosage; I had to take the low-est effective dose while maintaining the target international normalized ratio, or INR, which measures how thin blood can get, to see how long it takes blood to clot. I had my blood drawn several times a week, to measure and then maintain my INR.

Someone who has no clotting problems should have an INR of approximately 1. The higher one's INR, the longer it takes blood to clot. As INR increases, the thinner blood is, and the higher the risk of bleeding and bleeding-related events.

The challenge with warfarin is that INR can fluctuate—

due to diet and other circumstances. Regular INR measurements are needed. Even so, I bled. And bled. And bled. My periods were phenomenal.

When I cut my finger, I would watch the blood ooze for longer than I'd ever seen it do. Or I would stub a toe or bang my knee against a table and a beautiful bruise would blossom, larger than expected. I knew then that I was taking too much warfarin—the doctors would inevitably adjust my dose down. But I was a forgetful patient, and thus a bad patient. I loved seaweed, which contains vitamin K, which thickens blood. Because I ate so much of it without reporting, the doctors increased my Coumadin dose. But I didn't eat it every day, so when I stopped eating it, the new Coumadin dose would then cause my blood to thin so much that I had trouble clotting. I developed track marks on my arms from the blood draws. A cut was not just a cut, but thin blood was better than clots, was better than a potential stroke.

At home I received Lovenox shots as part of a bridge protocol until the warfarin kicked in. I was supposed to self-administer the shots, but I could not. Before the stroke, I was the kind of person who would sterilize a safety pin in the fire of my stove and then poke around for a splinter under my skin. After the stroke, I sobbed every time I thought about shooting myself with a needle. I opened the box, unwrapped the syringe, and squeezed my belly fat, but I could not inject myself. So Adam did.

Adam had to chase me and lure me toward the Lovenox, all while I screamed, "No shot! No shot!" before he pinched a piece of skin on my abdomen and injected the drug each night. The shot hurt going in, and the injection site stung like a bee sting for half an hour afterward, even when vigorously rubbed, per my doctor's recommendation. I cried hysterically like a toddler, unable to control my fear throughout. I was unable to console myself.

Years later, when my daughter became a toddler, I would recognize this same panicked and distraught behavior in her when I had to treat her asthma with a mask inhaler.

Each shot produced a bruise the size of a fist, green and purple. My abdomen was littered with these polka dots. On the last day of Lovenox shots, I did a happy dance.

Later in my recovery, when I was well enough to understand what had happened to me, to realize my deficits and become depressed about my stalled progress, when I wondered if my old life would ever return in any familiar form, I pondered taking a big dose of warfarin and then slicing my wrists.

———

One of the things I had to do once out of the hospital was to repeatedly explain to visitors and coworkers what it felt like to be in my brain and my body.

"How are you feeling? What's it like? But you look fine— did you really have a stroke?"

"I really did have a stroke." I could not even access my normal sense of sarcasm yet.

"I really can't tell that you've had a stroke."

"I can tell," I said.

"I'm not lying."

It really did happen. I really was not well. I was already

quiet and I became quieter. People excised me from their lives—those at work previously dependent on me stopped calling, and friends ill at ease with what had happened to me went away.

But I wanted to tell the world what was happening. I blogged under a pseudonym. And there it was that I wrote every single day. It took me hours to write one blog post, and sometimes the post itself made no sense, but I did it anyway, because it was something to do and something I wanted to do and something that told the world that I was still there. I am still here. I wrote down the words that came into my head in the order they appeared. My internal editor was turned off. I wanted to scream out into the void, but eloquent words were gone, so what I screamed was guttural and primal and, it turned out, my truest emotions, the ones unedited and un-filtered by my frontal brain, untouched by judgment and or-dered thinking. It was my body speaking.

In February of 2007 I wrote:

I'm in emotional and mental pain. It's awful.

I'm getting better much more slowly—the first dramatic weeks of recovery behind me. Now I'm looking at a mountain peak hundreds of feet up, with slower progress, less emotional resolve, and a broken leg. Oh, and a broken leg that no one can actually see, so I don't get a lot of forgiveness for my deficits and instead get a lot of impatience and quizzical looks. It's a very lonely road.

It's up to me to say, "No, I can't do that," and that's an overwhelming task for me to accomplish. Recognize and voice my limitations? Sheeit. And never mind the people who like to say, "You don't LOOK like you've had a stroke!" Sheeit. I feel like a stupid person who can't remember things well, and this all sucks now in a frightening way.

I cannot hold my thoughts and I'm sick of being a receptacle of only the present moment. (I watched Finding Nemo *last night and actually empathized with Dory, the fish who is dealing with short term memory loss—it was both funny and horrifying). The lesson of living in the moment is wearing thin.*

My brain was the part of me I always relied on most. I don't like the way I look and I am so not an athlete—so I always relied on my ability to THINK and REMEMBER and rely on my BRAIN.

And now it is GONE. I'm infuriated and depressed. And will I ever be the same again? No.

I am not driving the bus. Someone else is driving the bus and I've got to follow my brain's pace. I want to go back to work, but I cannot. I want to remember things, but I cannot. I guess it's time for me to learn a new lesson: patience.

All day, every day, my state of mind was most equivalent to the fog of having just woken up but not yet being fully awake. This state has an official name, sleep inertia, during which time frontal lobe function is depressed in the first few minutes after waking, slowing reaction time and impairing short-term memory. Damage to my dorsomedial thalamic nucleus impaired that very part of my brain. Sleepers often wake confused, or speak without making any sense, and

often want to go back to sleep. This is also called the hyp-
nopompic state, a term coined by the psychical researcher
Frederic Myers.

I spent the first few months of recovery in this groggy state
of mind. Except I didn't appear absent—my eyes were open,
my hair was groomed, my teeth were brushed, I was not wearing pajamas, and I appeared to be engaged in conversation.
But my mind was not fully awake.

I would wake up and open my eyes. I would wait for my
mind to clear, for the events of the day prior to flood my mind,
for the day ahead to come into sharp focus, for the ability to
envision my concrete plans, and for the ability to hold a cogent
conversation. I would wait for my energy to enter my body, to
feel invigorated.

But those moments did not come. Where was I? I must be
at home. The surroundings looked familiar.

Eventually, I would rouse myself and climb out of bed. I
put on my clothing for the day without purpose, for I could
not recall what I would be doing later. So rather than choose,
I wore mostly T-shirts and jeans, two things I knew could
not be mismatched, and which I could grab and don without
much thought. Choosing an outfit was a near impossibility, a
major exercise.

I brushed my teeth.

I petted my dog.

And then I sat.

I sat for hours. Or maybe minutes. I was unstuck in time.

Like Billy Pilgrim in *Slaughterhouse-Five*, I experienced my
life like a random series of moments without beginning or
end. All points in time existed simultaneously—and for me,

that was in the present moment. Nothing happened before, because I could not recall the past, and nothing happened after, because I could not comprehend the future.

I ate lunch but did not remember what it was I ate. It was lasagna and spaghetti and bibimbap and hummus and crackers all at once. I walked into the kitchen, and then it became the living room became the bedroom. I had no control or insight over where I was going next. I never knew where I would end up. Who and what I would see.

———

I was born in 1973 in Queens, New York, the eldest child of my immigrant parents, who had survived a war, who had gone to the top university in Seoul, who were recruited by the United States as an engineer and a nurse, who landed in postapocalyptic 1968 New York City. I was a chubby baby who my father described "not beautiful like Elizabeth Taylor, but very smart." We moved to California. I graduated from my suburban California high school in the top 3 percent of my class and attended UC Berkeley. I was supposed to be premed, but I purposely flunked organic chemistry, so I graduated with a degree in English literature. I lacked the courage at the time to tell my parents what it was I really wanted to do. And so I created an obstacle instead. And life gave me an obstacle in return. Which made me the writer I became.

I got married, or rather, I eloped with Adam, after college. Our parents were surprised. They demanded a wedding. So we had a wedding ceremony. Two, in fact. A Korean wedding and a Jewish wedding. Bowing. Seven circles. A rabbi. *Kippot* for all. The hora. And then years later I had a stroke. *So it goes.*

I did have thoughts after my stroke. But they were fleeting and sprawling and I could not keep hold of them. My prefrontal cortex was unleashed and traveling through time as I sat on the couch in a daze.

I went to sleep on the couch a coherent woman and traveled back in time to my wedding day, then forward to the day of my daughter's birth, then forward again to the day my husband asked me for a divorce. I walked through a door in 2006 and came out another one in 2007. I went back through that door to find myself in 1996. I was studying with Rabbi Finkelman, an Orthodox rabbi who did not ask me questions but directed me to ask questions. I was discussing *The Kuzari* with him. I was spastic in time. I lost track of time. I lost the ability to string my memories together in one cohesive narrative. To extract themes from that string. My mind and memories wandered, regardless of time and space.

Like Billy Pilgrim, I'd broken my head open. And I felt the truth pour out of me, which was all the emotion I'd learned to manage by bottling it up or shelving it. Because my head was broken, I cried upon feeling hurt, and I raged upon feeling indignant. I could not stop myself.

When I finally got home after my hospital stay, I was quiet for a while in our quiet house. Adam worked from home for three weeks, making sure I ate and showered and otherwise met my basic needs. While I slept, Adam ripped up the wall-to-wall carpets in our house. He was a peripatetic person suddenly confined to our home. And so he created projects for himself, and our home began to change in appearance.

Underneath the carpet were dark walnut floors, original to the house, untouched for more than fifty years. I admired the grooves in the wood and the darkened nails.

The floor had been there beneath us all along, for years.

It was new to us.

But it was not really new.

He painted the bedrooms, too.

And so on.

───────

I continued to check in with work, at our start-up where I ran the human resources department. I worked on autopilot. I showed up for a few hours each week to make sure the most crucial tasks were completed, and then I went home, utterly depleted.

I was lucky that my boss tried to understand what I was going through. He tried to take everything off my plate so that I had the minimum of work tasks—ensuring that payroll went through, that benefits were active—and then I clung to these tasks in protest, as if the act of helping me was the most unjust act of all. I was in denial, and I felt lessened by his act of lessening my workload. He was the CEO. And as the CEO, he told me, "You're either sick or well. You told me you're sick, and so I'm taking work off your plate."

I sat in his office and cried in hysterics. On his desk was a dried-flower arrangement, one that I'd given him a few months earlier. No one else at the office had given him flowers on his company anniversary. I liked him. But at the moment I did not like him.

"Christine," he added, "you are being moody and mercurial."

"I am not!" I'd never screamed at a boss before.

His ramrod-straight spine, a product of his Andover education, straightened out even more. He went silent. Not in anger, I believe, but in confusion and alarm. This stark contrast, between my unhinged fury and his calm, between the diplomatic demands of my work role and my loss of control, made me quiet, made me pick up my notepad and leave his office.

In hindsight, I am horrified by my mood swings at work back then. It's not that this was the only time I was horrified by my own behavior during recovery. It's that I had a tantrum in the workplace. It's that I was the director of human resources and I'd lost my cool. It's that I no longer had any sort of emotional barrier in the workplace. It's that I'd cried multiple times in front of my peers, when before it was my rule that if I had to cry, I would never do it on-site. It was that I'd never have yelled at a boss. It's that I'd violated my own personal laws. It's that I couldn't help but do it.

I could drive, even if I did not understand directions. Even if I forgot where it was I had parked the car. Several times a week I got myself to the lab to get my blood drawn so that my INR could be measured. I am not sure how it is I got there each time. But I got there. And then I got home. I moved from space to space without awareness, without a sense of time.

And I didn't mind. I wasn't aware of my deficits. These things just happened.

My brain was healing, though. And quickly, in the first six months.

7

Caregivers play a significant role in stroke recovery, starting from day one. The nurses at the hospital made sure to monitor me. To record all facts and happenings. To put me at ease. To facilitate my healing. I am indebted to their hard work and compassion. But I did not know them. They did not know me. Nothing I said or did was emotionally charged. They did not have to live with me forever. They went home to their own families each night. This is what they did every day, what they went to school for, what they receive a paycheck for.

When it is for someone you know, caregiving can be a monumental challenge. It is described as the most difficult job for which no one has applied. It is an unexpected and unwelcome job. An unpaid job on top of another job. A job caring for a beloved who is no longer the same, who is short-tempered and demanding in new ways. Who no longer resembles the person you love. A job that takes a physical, mental, and emotional toll. A caregiver has lost a source of support while taking on the physical burden of feeding and cleaning and providing medication to that very former support, alongside grief, which must be set aside. It calls for strength to take care of a loved one, and it presses upon the caregiver's weakest seams, too.

Adam was my caregiver.

And he took care of me well during my recovery.

I can't even recall all that he did, and that makes me feel sad. I couldn't be there for him in the ways that I had been, and that makes me sad, too.

Adam was nineteen and I was twenty-one when we began dating. He was tall and serious, with an olive-skinned face that lacked baby fat or mirth. He could not drink legally but would never get carded at bars because of his visage and height. He worked on machines that did not care whether or not he laughed or cried or endeared himself to them. The computer science courses he undertook at Berkeley cared only about accuracy and efficiency of code, and so did Adam.

"Have you ever been depressed?" I once asked Adam.

"No," he said.

"Why?" I asked.

"Why not?"

We chatted online for months, before there was AIM or Facebook Messenger or SMS, on an archaic UNIX chat system at Cal that Twitter resembles. And then we met in real life.

This boy named Adam bragged about his Camaro. I mocked him in return. He had to meet me, he said. Prove to me he had a Camaro for real. That it was green with rocking bass from a bazooka. I said I didn't care. He said he did care. He wanted to show it was true.

CRISTINE> you don't really have a Camaro.

ADAM> yes I do! I will prove it to you.

CRISTINE> okay. But I don't even care about cars.

ADAM> How can you not care about cars? I have to meet you. I'll change your mind.

CRISTINE> I don't want to meet you.

ADAM> I'm going to keep asking until you do.

Three months later I went out to dinner with Adam. I opened my apartment door to a tall, thin, dark-haired boy in a button-down shirt and burgundy jeans. There was something in the way he moved, the way he put his hand to his chin and then back down to his side again that made me recognize him: two years earlier, I'd been on my way to an early-morning class when I decided to check my e-mail at a campus computer lab. A boy crawled out from beneath the desk. I shrieked and ran.

It was Adam, I realized, who was the boy who slept under the desk in the computer lab. Who had stayed up all night keeping the servers up, and then nursing them so they kept running. Who had then spooked an undergraduate English major from her computer.

"I think I know you," I said.

"Yes. You agreed to this date."

"No, that's not what I mean. Wait—is this a date?"

He smiled. That wry grin, one that someone who has not yet gotten to know him might miss, so subtle it was, the outer corners of his mouth only just tucked in. "What else could this be?"

I sat in his Camaro that evening in a black velvet skirt

and brown Rockport sandals and made fun of his car. He showed me the stereo. I stared out the tinted windows. It was true. The car existed. And then it was done, I thought.

We went to a Burmese restaurant, where I ordered all vegetarian dishes. Weeks later I would learn that he went home afterward and with his roommate chowed down on Quarter Pounders from McDonald's.

"You did that?"

"Yes," he said.

"Why?" I asked.

"Because you don't eat meat."

I was touched that he accommodated my needs. I was stunned that he didn't voice his. Later, when he became my caregiver, he experienced sacrifice again.

I don't know why he fell in love with me. I wasn't particularly charming to him. I didn't even want to be in a relationship. I was a mess. Maybe he saw something in me that was missing in him, like I did him.

We were in love in the way young people fall in love, in an all-consuming, self-forgetting way. I forgot that I was an avid coffee drinker and I didn't drink my morning coffee, because I was with Adam and he was enough and I barely remembered to go to work, let alone make coffee with a machine that did not exist at Adam's apartment where we spent every night, because my own apartment was too much of a hovel and the bed I owned was a used and donated, lumpy futon. Adam didn't drink coffee, and thus I didn't, either. I had a headache the entire first three months of our relationship. I woke up each morning with a pounding head and then I met Adam and forgot my pounding head. Until after months, I no longer experienced a pounding head.

I wanted to adopt his way of life. I studied with an Orthodox rabbi for five years with the intent of converting. *Baruch atah Adonai* before dinner. *Chag sameach* on holidays. I wore a hamsah around my neck. I stopped eating vegetarian and ate more meat in one month than I'd eaten in five years. I could not differentiate one Porsche from another before I met Adam. A Porsche was a Porsche, mostly because of the name engraved on the back of the vehicle. The difference was the color of the paint. So Adam took me on midnight dates to car dealerships, showing me the different Porsche models one after another: Carrera, 911, 928, 944, 914. This, until I could name them all, albeit with reluctance.

He, who, I would learn, did not read literary fiction on any sort of regular, let alone occasional basis, who I never saw buy a literary book again, bought a Milan Kundera novel and read it alongside me. I asked him to read my copy of *The Great Gatsby*, and he did. He learned about Daisy and Gatsby and Nick. Adam discussed Gatsby's car. Was it a Rolls-Royce? Or was it a Duesenberg?

We were, in those early days and weeks and years of love, bending ourselves toward each other. We were hungry for each other. So different we were, and still are. And we met at the fulcrum—I, with my intuitive and emotion-driven style, and he, with his clarity of thought and even temperament. I could detect the emotional temperature of a room, understand who it was my allies were. But Adam could put two seemingly disparate ideas together, with the countenance of Spock. He could engineer anything. Together, we were a jalopy of coordinated strengths.

He was the most brilliant boy I'd ever met, one whose ambition and drive were unmatched. He had grand visions for his life that I both envied and found lacking in mine. He wanted to be a technical revolutionary. He wanted to found and run a start-up. And he did so, eventually founding a big data software company, before becoming a venture capitalist at the top of the high-tech ecosystem a decade after that. He achieved his dreams, while I was too afraid to pursue mine, which was, at the time, to write and publish fiction.

That someone had such confidence in himself was amazing to me. Such certainty.

I met him the autumn after I tried to kill myself for the second time. I no longer wanted to live the life I'd been living, the one in which I tried so hard to be perfect and failed. I had not gotten straight As. I had not gotten into any graduate programs. My life as a student, the only life I'd known, had ended. I had brutalized men, and in turn they had brutalized me. I held the skin of my perfect self. Inside that skin, I was air. I was nothing. I was dissipating.

I was not jumping through time. I was not jumping through anything. I was stuck.

I had graduated from college and was not utilizing my English degree. I took a job in high-tech as a UNIX system administrator and then as a recruiter, because those were the only jobs available to me. My life was disjunctive; what I did before was not what I was doing, what I was doing had nothing to do with my life before.

So I spent an entire month locked inside my apartment. On a couch. I cut my wrists. Every day. Shallow cuts, as I had in high school. To punish myself. To inflict pain on the body in which I was trapped, which brought me so much psychic pain. To, in turn, feel pain. To feel anything but nothing.

And then I stared at Tylenol, at Advil, at NyQuil, at anything I could take too much of, a desire that grew stronger at sundown.

I tied myself to a chair each evening, scared of what I would become and do. Like a werewolf. Like I was a monster.

I made that apartment my prison. And I died a little each day, until I decided to change my life or die.

I wanted a common thread in my life. This was around the time I started corresponding with Adam.

During my month in the apartment, I went on the Internet, which back then was a terminal screen and command line prompts. Telnet. Talk. Write.

Four months after our first date, Adam asked me to marry him. I said yes.

My life became a continuous line. I had met Adam. And I would be with Adam forever. I became his fiancée.

And then I became his estranged fiancée when he broke off the engagement three years later. And then I became his girlfriend when said he wanted me back a year after our breakup, after I'd adopted a wiener dog and bought a house but before I'd gotten over him. And then I became his wife.

Becoming a caregiver at the age of thirty-one for one's thirty-three-year-old wife is a disruptive change.

I had become, at the age of thirty-three, unable to balance the checkbook. The numbers were the numbers on the page, but they turned into shapes and squiggles when I tried to put them together. When I put them together, I made new shapes and squiggles. Those shapes and squiggles were apparently not the correct sum. Adam asked to see the checkbook, and because he had no other way to react, he scolded me. "This isn't math," he said.

"It is math," I said.

"You aren't adding and subtracting correctly."

"Yes, I am."

"No."

"I thought I was."

I was unable to manage any household tasks like calling a plumber and then telling the plumber what it was that needed to be fixed. That the shower would not stop dripping and Adam had tried changing this thing or that and tightened the—and I would forget. My thoughts would spin, and I would feel overwhelmed by the time I dialed the plumber's number.

"What's the problem?" the plumber's assistant would ask.

What was the problem? The problem was the shower and the plumbing. The problem was somewhere upstairs. The problem was something not connecting. The problem was that there was a disconnect. The problem was something had broken. There was a problem. And then I would hang up.

Instead of Adam and I being equal partners, I was an emotional and mental invalid.

It was assumed that Adam would do things for me because I could not.

Particularly deceptive was the lack of physical symptoms, and how I appeared to be well but was not well. My hands worked. My eyes worked. My legs worked. Nothing looked broken on the outside. No scars. No limp. But my brain—my brain did not work. I could not think. I could not hold thoughts. I could not remember. I became infuriated at my lack of memory. I raged at the lost thoughts. I cried. I tried to explain what was wrong. But I could not explain things. I threw things. I cursed. I was a toddler. I looked insane. It was easy to write me off.

And then I became so afraid of doing anything, I insisted on doing nothing. When even after two years Adam asked me to coordinate household tasks, I would flinch and refuse. I was scared. I knew I was better, but I was overwhelmed. I wanted to do other things. I had limited energy and wanted to allocate what resources I had to the things I felt had value to me, for the things I wanted to do. Life had become short.

The person to whom I turned to achieve these tasks was Adam.

He could not confide in me anymore. He could no longer solicit advice from me. He could not rely on me to take on my portion of the small household tasks we once shared, like paying bills, making meals, taking our dogs for veterinary appointments, or making dinner reservations. I could not even remember to take my medication, and he had to make sure I did that, too. When he went back to work, he had to call me several times a day to make sure I had not gotten lost or burned the house down or fainted from having forgotten to eat.

He sat next to me while I wept. He stood next to me in line at the pharmacy while I raged. He watched his composed wife break down. He could not advise me on how to better manage myself, because how would I remember?

Throughout my recovery, he helped me without complaint. Let me repeat: he helped me without complaint.

There was so much in his silence, though.

He wondered if I would be disabled forever. He asked himself, "Will I be doing this forever?" I know this, because he told me years and years later that this was running through his mind. That it felt like panic. That it felt like prison. Like a weight on his shoulders. Like doom. That, I suspect, it felt like a silent scream. I had no idea. He sucked it up and absorbed my tantrums and my depression and my dependency.

He handled the checkbook. He laid out clothes for me to wear. He made financial decisions. He reminded me to eat. He helped me decide what to eat. He did this for years.

And because I was not gracious in those months following the stroke, caregiving was also a thankless job. In turn, Adam became embittered. Looking back years later, he told me, "I felt you wanted special treatment. I think I resented you and thought of it as bullshit when your cognitive abilities seemed to be there but you used excuses and crutches all the time."

Our silences destroyed us.

But then again, how could Adam even articulate this to someone with brain damage in a way that she could understand?

Stroke recovery is not a linear process.

A relationship is not linear, either. It lurches forward, stalls, backpedals, steamrolls forward, leaps, dances, falls. It can take different directions. A crisis can put a relationship into a tailspin. Or maybe a marriage can float but be struck by an iceberg. Some marriages can grow closer in the wake of illness. And some relationships break. All relationships change. Because illness changes a person and the dynamic of relationships.

We were no longer a team in the way we used to be. No longer would we divvy up tasks. No longer would I undertake them in their entirety. Not for at least two years in what was then an eight-year-old marriage.

A part of me had died. I had become different. I was adjusting to my own landscape, and transitioning to sickness and then back to wellness took its toll. And unbeknownst to us at the time, it was eating away at our marriage. It was eroding its base.

But then the stroke happened. And I told myself I would live every year as if it were my last. And in doing so, I stopped living under Adam's tenets. I rebelled against the structure of my life, and because that structure was Adam, I rebelled against him, against his family.

Here is how I lost Adam. Here is how Adam lost me.

Here is how I lost myself. Here is how I found myself again.

8

I was home.

I sat on the couch. I stared at the wall. And when I didn't stare at the wall, I watched a lot of Food Network television. Ina Garten walked through her garden to pluck parsley. Paula Deen cackled as she added cream and butter to her dishes. Giada de Laurentiis made pasta and bared all her teeth. Their ingredients went into pristine pots. Ina's into her dune-colored Le Creuset. Paula's into her cast iron skillet. Giada's into her stainless steel. I was hungry but could not register hunger—I could not translate my light-headed nausea into a need for food. What I saw on the screen had nothing to do with my hunger.

I let my mind drift, a boat on a wide sea without anchor or sail or engine.

I noticed things. The sweet smell in the air as spring approached. I did not think the sweet smell was flowers. I saw it as a separate thing, in and of itself. And then it passed from my mind. My dachshunds sat next to me on the sofa. They smelled like Fritos.

I did not notice morning pass into afternoon. I did not know for how many hours or minutes I sat. When I looked at the clock, I noted the time: three o'clock. But it was just the time, because I did not remember at what time I had sat down in the first place.

Ideas floated into my head, then left, like waves on a river. They did not come back like ocean waves. I would have ideas for stories that I'd forget if I didn't jot them down in my Moleskine. Later I would stare at my journal and wonder about these ideas. Why were they there? To what were they connected? The ideas were one or two words: "Chickens" and "cotton dream" and "stormy lake." They made no sense. They triggered nothing. There was no context. Like dreams—I had woken up and forgotten them all.

I tried. I tried to hold on to ideas and thoughts. But they spun out into the ether.

My mind was a sieve through which so much fell. I could hold things for a second and see the shape of the thing the memory the idea the experience the person the feeling the flavor the warmth, but then it was like looking at a sieve through which I'd just sifted flour—I would only be left with the impression, the knowledge that something had been there that was now gone.

I'd take a sip of a drink and move on. I'd open a can of apple juice and forget. Then open another can. Fill a glass with water. By evening, when Adam came home, there would be so many opened cans of juice in the fridge and half-filled glasses of juice and water around the house.

"Are these all full?" he asked.

"I don't know."

"Honey, these are all full," he said.

"I'm sorry," I replied.

"Where are you going?" Adam would come toward me.

"I'm running away," I said.

"Because I pointed out the glasses?"

"Yes," I said. "I'm ashamed!"

"Why are you crying?" He had had a long day.

"I don't know!"

I cried all the time. I hardly recognized myself. I could not deflect and protect myself from any of life's wounds. I began to realize I should have known those glasses were full. Why hadn't I known they were full?

I still forgot to eat.

I could not connect what it was I saw with what it was I was supposed to do, with what the possibilities might be, with how to organize potential.

I was alone most of the day. And I wasn't lonely. I was with myself.

I had appointments, which I kept only because I wrote them down. I went to occupational therapy. I wondered why on earth I was prescribed career counseling. In my journal I noted that I spoke with someone in my occupational therapist's waiting room. She and I were both there for short-term memory problems. Mine because of my stroke. Hers because of electroshock therapy.

I have no memory of this.

I wish I did.

I lived entirely in the present tense, unleashed from the part of my brain capable of worry and anxiety and fear and strategy and power. I was happy and peaceful. Nothing was in my control, and I did not care. I opened up my wings and flew with the wind. I fluttered down from a tree and traveled on the surface of a stream, the veins of my leaf body bejeweled by water. I traveled about a room, a dust mote lifted by the gust of an opening door or a fluttering curtain, lit up by the morning sun. I feel those first weeks of stroke recovery as a series of unfettered, perfect moments, the kind that I now seek through retreat or in yoga or music.

I am filled with nostalgia for that period of time; they were a gift to me, a state of mind that so many people seek in their adult lives.

I was brain-dead. So disabled, yet blissed out. This, in stark contrast with what was to come—a heightened awareness of my shortcomings, a darkness to counter the lightness of those early weeks of recovery.

What was wonderful for me in those first weeks was not wonderful for anyone else. I did not care about things in an active way, because I was unable to. I did not feed my dogs until they clamored for food. I did not care about my marriage, so I did not make sure Adam went to work with a smile and went to sleep without stress or worry and I did not check in with him to see how he was doing and I did not ask him about his day and I did not remember what his days were like and I did not do the laundry and I did not cook meals and I did not console him when he lost a business deal and I did not encourage him when he was about to go onstage to speak.

I did not care about my friendships. I did not support friends who were going through hard times, and I did not care about their successes, because I was unable to hold these happenings in my brain, even if I did love my friends and I loved my husband and I loved my dogs.

I was barely aware of my own needs, acting only because my body told me I needed to use the bathroom and made me dizzy and light-headed when I did not eat. It made me cry when my feelings were hurt and it made me rage when my needs were unmet.

All these things had consequences.

My body learned to be heard. My brain realized that my body had value. My mind wove stories and learned to incorporate my body and my brain in its decisions and narrative.

All these things had consequences.

My life changed vector, if even slightly, perhaps one degree off original course. This one degree was undetectable at the time. My marriage did not feel it just yet. I did not feel it just yet, either. But I was learning to be at peace. To focus on myself. I was learning who my real friends were. That sickness can happen at any time. That life can end. That our abilities can be taken away at any time. That all I wanted to do was write again.

All these things had consequences.

Over time, even a one-degree shift can take you way off course.

The future is the past is the present is the future is the past. I am time traveling again and again. I am in the hospital in that room, and no one can tell I am not well until there is proof. I am out of the hospital on the curb and the world is loud and bright and I am small and trying to hold very still. Adam is holding my hand and then he is not holding my hand and then he is holding my hand again. I have had a stroke. I have survived my stroke. And it is behind me and yet ahead of me around me and the lessons from the stroke never end and the things that end come back again.

I am alive but other people die. And when they die, I will be reminded that I could have died but did not die. My eighty-three-year-old father will have a major stroke seven years after my stroke and survive but will never be the same again. He will tell me he has already priced out his memorial service and double-checked on his burial plot. He tells me all his friends have died so the guests will be my friends, my brother's friends, and my mother's friends, and I will suppress a split-second desire to tell him to stop just please stop but by then I will understand what it means to have had a stroke and to want to be prepared for something that could happen at any minute and I will nod when he gives me directions and I will promise to visit his grave each year and I will thank him for his thoughtfulness and I will tell him I will do as he says. My forty-six-year-old friend Justin will have a massive stroke and be taken off life support a week later and then linger as if on a precipice before falling transitioning saying good-bye without words even though he was a poet and words were his life he dies in silence except for the monitors like an exclamation point. And what is left of him are memories. And I know that memories fade, memories warp, memories die with you, too.

Though my stroke happened in the past, it is my world today. I know that everything is temporary and hard to hold, and
even the permanent is only for this lifetime. That the poem
Justin wrote for my newborn daughter on a piece of paper will
eventually yellow and crisp. That all the secret family recipes
of the world will disappear unless we share them. That all this
value and love and honor and pain is useless until we share
it with others, so that they too can hold it in their hands for a
split second before the dust runs through their fingers.

<hr />

Each day was an hour was eight hours was twelve hours was
another day was a week was all running together in time. I
was fed. There was a bed. There was a couch.

My mother wanted to visit, but I did not allow her to. I did
not want her to be sad. I could not handle sad. I wanted everything to be predictable, scheduled, reliable, and emotionally
undemanding. I still do. It's hard for me to leave home. It's
hard for me to leave the familiar and adjust to new surroundings. It's difficult to adjust to a new schedule. Don't take me
wrong; I love new things and new places. But transitions of
any kind are difficult to undertake and require me to make
time for rest and self-care.

I underestimated my mom's ability as a nurse—namely,
her ability to compartmentalize her feelings; the thing that
was a challenge in all other areas of her life was the thing that
enabled her to care for others with excellence. She knew better
than anyone how to not burden the sick. But for now, I had

forgotten that. For now, all I wanted was to be alone in a house that I knew so well, I could navigate it with my eyes closed.

Adam had to take a business trip three months into my recovery. Because I couldn't be left home alone, because I could not remember to eat and when I wanted to eat, because I could not figure out how to assemble food to feed myself and because I might set the house on fire, I accompanied him to Las Vegas for a conference. I followed him around the airport with my ears covered against the din, and I stayed in the hotel room where I lay in bed listening to the gentle hum of the air conditioning. Adam checked in on me every couple of hours to make sure I was okay.

A day into the trip I was in the hotel room when I received a call from Adam. I picked up the phone as I'd done two times earlier that day.

"She's dead!" he shouted.

"What? Who?"

"My mom! She's dead!"

"Are you kidding?" I asked. I will always regret having reacted this way.

"No. She drove her car into Topanga Canyon."

I said I would pack the bags and meet him down in the lobby. That we would take the car and drive to Los Angeles, where we would meet his family.

I did this on autopilot, in the way that people in shock and denial and grief do. In the way that I had been living my life for the past few months. So it goes.

We drove to Los Angeles. I do not remember the drive. She had driven her car off a cliff. There had been a helicopter flown in to rescue her. She died on the scene from a broken neck. There was a coroner. There was a mortuary. There were arrangements for plane tickets. There were phone calls to El Al on transporting her body to Israel. There were phone calls to friends and family. There was crying. I held my husband's hand while we sat on the bathroom floor and he cried. There was her closet, clothes hanging empty of her body yet redolent with her perfume. There was my sister-in-law, burying her face in her mother's clothing. There was the last thing she had eaten, a piece of cheese curled from heat and neglect on a plate.

We flew to Tel Aviv to bury her. I cried at the airport ticket counter when the ticket agent upgraded us to business class. I cried in the plane. A man in the aisle seat would not trade his aisle seat for my aisle seat so that Adam and I could sit together. He would not. We told him our mother had died. He would not.

We escorted her body on the plane, in compliance with El Al's safeguard against terrorism and in the spirit of Jewish law, where as a sign of respect, a body must never be left alone until after burial. My mother-in-law was in the cargo hold below.

A baby cried the entire flight, choreographed to our grief.

In Tel Aviv we buried my mother-in-law next to her mother and her father. Adam and I drove straight from the airport to the cemetery, changing into our funeral outfits in the taxi. The cemetery looked like a desert because it was in the desert. There was no grass. It was an old cemetery filled with gravestones, flush against one another. Cousins who were *Kohanim* lingered outside the gates of the cemetery. My father-in-law, sister-in-law, brother-in-law, Adam, and I walked in. We tore our clothing. We could not wear shoes that were either leather or new for a burial, so we each wore a pair of black Crocs.

We stood in the sun, still warm in the winter desert. And then they brought her out.

I expected a coffin, most likely a plain wooden coffin, but they brought her body out in a white shroud. It looked like a straitjacket that also covered her head. There was a long pole inside the shroud along with her body. She had to be buried with objects covered with her blood. So it was a bloody pole. Or what must have been something resembling it. From where had that long rail come? I didn't ask to look under the shroud. I didn't want to know the terrible fact of which car part or what utility pole contained so much of her body that it had to go into the ground with her.

I expected a coffin, but what I saw was her body. They flung it into the grave. It felt like she was a piece of rubbish. She was going into the ground. And then the rabbi, whom we had never met before this day, said the prayers. *Baruch atah Adonai.* A rooster crowed nearby. The soil was shoveled upon her body. I'd expected a coffin, and here was her body, going straight into the ground. She was covered in soil. Her feet, her arms, her torso, her head. The outline of her body in the ground. I heard someone screaming and crying. The rooster stopped crowing.

Later, after we returned home to Berkeley, I told Adam, "Your sister—she was crying so hard."

He looked at me. "Honey, that was you."

I couldn't keep my emotions in. My brain was too damaged to control or manage my grief, and I wept from an immediate, sincere, and intense place. To express my emotions as I felt them—this was new.

We sat shiva in a hotel after the burial.

Adam and I slept in a room with curtains useless against the light reflecting off the Mediterranean Sea. We didn't even think about which side of the bed we each slept on—one night Adam was on the left, the next he was on the right.

I could not adjust to the time zone. I awoke in the middle of the night. Slept in the afternoon. I came down with the flu. Shivered in bed with a fever. Relatives came by with food for shiva. We were not to leave. We were to be cared for. *Mahalabia.* Pita and hummus. Falafel. I snuck out once each day once my fever broke. To the market. Shouk HaCarmel. I brought back Iraqi nougat, *Baba kadrasi.* Strawberries. Baklava. Bissli snacks.

My memories of the time are jumbled and jangled.

Vonnegut wrote, "There is nothing intelligent to say about a massacre."

I write, There is nothing intelligent to say about a death.

Everybody is supposed to sit shiva and remember and be fed and share memories. I wanted to be forgiven.

> *Everything is supposed to be very quiet after a masssacre,*
> *and it always is, except for the birds.*

A rooster. And what does the rooster say except "cock a doodle doo"?

Poo-tee-weet.

So it goes.

―――――――

I wished I'd died instead of her. While sitting shiva in Israel, surrounded by grieving family and friends, I wanted to console everyone by dying in her place. I felt half dead, anyway.

I had not spoken to my mother-in-law in three years by the time she died. We had been angry with each other for that long. I said sorry for your loss, sorry for the situation, sorry for the silence. I felt my life had changed forever. In my journal I wrote from Israel:

> *I am not prepared for change. Even now, I feel my marriage*
> *and Adam, changing before me. There is a new distance and*
> *it scares me to death. He is slipping out of the familiar. The*
> *other night he saw me upset and yelled, "You act like you don't*
> *want to be with me anymore!" That was how he read my grief.*
> *Our lives have changed forever. And I don't like it.*

I could not remember much while sitting shiva, even as the rest of the family shared memories. My memory came into sharper focus later, as everyone else's memories faded. The way her voice sounded. The way in which we'd parted. "The next time we see each other," I said, "one of us will be dead and going into the ground." I'd said it because she'd betrayed me. Because I was unwilling to forgive.

Those were my last words to her.

How true.

How awful.

How she was dead and I was alive. Or really, half alive. With no promise that I'd be alive again.

This grief was overwhelming.

I lost my husband to his grief.

And then I was left with mine.

I began to remember what I'd been able to do before the stroke. And because I could remember, I became aware of my deficits. I became aware of my incompetence. My inabilities. The bliss I'd experienced in the first few weeks of recovery was now long gone. I saw a great divide between the old me and the current me. I would never be the same again.

My grieving dovetailed with my stroke, with shifts in my my life, with changes in my marriage.

And what I saw, all I felt, was loss.

Grief is a natural part of recovery from any life-changing injury or medical condition. This makes sense—the survivor is mourning lost abilities and lost time and lost friends and lost identity. It also makes sense that the survivor does not take grief into consideration as a part of recovery; after all, shouldn't one be grateful for simply being alive? For having survived?

But no. It did not work this way for me.

I was angry and sad. I was not yet depressed. Depression would come later. When I would succumb to helplessness. When I wanted to give up. When my anger took over everything, covered up my sadness, drove me forward, even into darkness. I mourned the loss of my mother-in-law, I mourned the juxtaposition of my near death with her death, and I mourned the loss of my abilities, the loss of my old self.

But even then, I see that in my journal I wrote:

I saw this movie called The Last Kiss *in the hotel in Israel. It sucked except for one line: "You can't fail if you don't give up."*

9

In order to function, my brain had to heal and change it-self. No other body part is both the thing that conducts healing and the object of the healing.

Wound healing in the body involves hemostasis, inflammation, repair, and remodeling. The brain, the conductor of healing, registers pain and injury, and makes sure the body constricts blood vessels and coordinates clotting and the formation of fibrin mesh through platelet plug formation and the release of various chemicals such as adenosine diphosphate, serotonin, thromboxane A2, and prothrombin. Once the wound is closed, the brain directs the clearing out of damaged and dead cells along with bacteria through inflammation; the brain makes sure to send white blood cells, which eat the debris, to the area. Once cleaned, the wound then repairs and remodels, again under direction of the brain. The skin regenerates. Collagen realigns.

Experiencing the healing of the thing that so closely affected my behavior and personality and thoughts was confusing and chaotic and exhausting. Things had gone haywire—I existed, and yet I did not, just as my brain existed and yet did not. I was going down the stairs, but my foot slipped because the stairs looked like a two-dimensional painting but they indeed are still stairs and my brain misses this and so then does my foot and I fall.

Healing is exhausting. Plasticity comes at a cost.

The brain, while it heals, does so at the expense of energy— I was always exhausted. I was still sleeping up to twenty hours a day.

Sleep affects plastic change by allowing us to consolidate learning and memory. When we learn a skill during the day, we will be better at it the next day if we have a good night's sleep.

Marcos Frank said that sleep enhances neuroplasticity during the critical period when most plastic change takes place. REM sleep also helps the hippocampus transform memories (especially emotional memories) of the day into long-term ones. It makes memories permanent, thereby causing structural change to the brain.

Infants spend many more hours in REM sleep than adults, and neuroplastic change is at a peak for babies.

I slept.

And I still could not remember.

———

I had to learn to prioritize my desires and needs, something that when fully recovered, I continued to do. I went to occupational therapy to help with recovery—and my memory started to come back in slow stages; from fifteen minutes, to an hour, to half a day.

I learned that I had a limited amount of energy each day. I had to prioritize my activities and goals. In the beginning, grooming and coordinating an outfit for the day meant no blog post later. A social appointment meant that I could do nothing else the rest of that day, and needed a slower day the next. And when I meant to do nothing, I meant nothing. Not surfing the web, not vegging out. It meant sleeping. Or if I was conscious, staring at the wall in a stupor.

What I could do at nine o'clock in the morning I could not do at twelve noon and nine o'clock at night, if I was even awake then. It was as though throughout the day there were invisible weights increasing in size and mass on my body. There were invisible boulders in my brain, around which I needed to navigate and could not always succeed, and the boulders were so hard to move that by the time I got done moving them to get to the other part of my brain to try to access the memory about the time I was a little girl and my father took me to the ocean, just him and me, I was too tired and the memory too heavy and too hidden and deep and I needed now to move the boulder again just to get back to the place from where I came.

Over the course of a few months I would learn to conserve my energy, to know that I had a finite amount of gasoline. There could be no two days of activity in a row—I could not have friends to visit on consecutive days. I could not hold dialogue for that long. I could not remember what it was they said, and I could not figure out what it was I should say. In an ideal world, all I wanted was for someone to sit next to me quietly.

Each time I try to remember my stroke, it's different.

As I write this, it is November 2015. When I had my stroke, I experienced what is called the ischemic penumbra. Professor Robert Knight, a neuroscientist at UC Berkeley, likened the phenomenon to a sprinkler system, when I sat with him to discuss my healing: "Think of a stroke like a lawn sprinkler system. The sprinklers pop up, and the lawn gets watered. Then one of the sprinklers clogs, and that area of the lawn doesn't get water. The edges, however, get water from another sprinkler. If that dry spot is not reversed in three to five hours, the lawn dies.

"This happening," he said, "is called the ischemic penumbra. The edges may or may not survive, depending on access to a sprinkler, or the blood vessel. Part of survival depends on there being enough blood coming back. And some cells are completely lost."

I imagined my brain as a lawn. Withered grass. Brown grass. Parched soil. Starving. Dying.

He then described plasticity, or the healing process for the brain, in terms of a UNIX network system. "UNIX routes telephone info from AT&T. The route from New York to Los Angeles is normally through St. Louis—but if the system's power fails, it routes through Chicago. What the brain used to do in one damaged area, it begins to do in other areas, reorganizing, and using circuits it didn't previously use."

I saw the grass in the watered areas flush with green against the dead spot.

———

New pathways are built when we learn a new language, when we encounter a new kind of music, or when we finish reading or writing a novel. It also occurs when there is injury to the brain—just as, if a meteor falls on a freeway interchange, detours must be found, and new roads built, as they were after the Loma Prieta earthquake, which caused the Cypress and Embarcadero Freeways to collapse. New neighborhoods and roads formed. Old neighborhoods, such as Hayes Valley, that had existed underneath the elevated freeway in eternal shade and darkness and noise blossomed and prospered. Eventually, there was a new network of roads around that scar upon the earth, like pathways are built around a scar in the brain.

The brain changes in response to behavior, environment, thinking, emotions, and bodily injury, whether that injury is blindness or a fractured arm or a stroke.

This is called neuroplasticity.

The brain's resilience is remarkable. Resilience is remarkable. In the spring of 2007, during the course of my stroke recovery, a tanker truck carrying 8,600 gallons of unleaded gasoline overturned and exploded at 3:42 A.M. near the eastern foot of the Bay Bridge—more specifically, the 80/880 interchange in the middle of the infamous MacArthur Maze. The heat from the flames was so intense that it melted the overpass and destroyed approaches to at least three eastbound freeways. The driver survived; he suffered second-degree

burns to his hands but was able to talk to the horrified workers at the sewage plant below the freeway. He then walked a
mile and a half to a gas station, where, in a state of shock, he
hailed a cab.

In the morning the concrete slab of freeway looked like a
blanket, draped without care. The fire was hot enough to melt
the steel frame, which then bent, which then took the roadbed
off its supports. The Caltrans workers, who were cleaning up
the mess, looked like tiny toy soldiers.

By the following day, the concrete had been removed and
the steel exposed. The steel was like something from a Salva-
dor Dalí painting, bent like a melted candle.

I wrote about this explosion in my journal—and at the
time I did not realize that in many ways it was a metaphor
for traumatic brain injury or stroke and ensuing plasticity;
there were so many things I did not realize at the time. For a
month the highway connector was closed as they repaired the
overpass. Whenever we had to drive to San Francisco, which
was pretty much every day, because that was where work
was, where errands needed to be done, we had to find different ways around the freeway and the ensuing traffic snarl.
Caltrans, the state transportation agency, set up temporary
detours, and drivers discovered further detours, taking side
streets and frontage roads to the bridge. Public transportation
stepped up its availability over the coming weeks. The traffic
was slow and tedious and full of uncertain commuters. But
over time the traffic sped up as people learned these new ways
around the closures. Commuters adjusted their work arrival
times, and many avoided peak traffic hours, learning when to
get on the road for quicker drive times. And many switched

to alternate transit options—Bay Area Rapid Transit (BART) posted record ridership numbers.

When the overpass was repaired I wrote:

> *It was an amazing feeling to travel a straight and direct path again, after weeks of meandering through surface streets or doubling back on the 80 . . . things are coming back to normal, aren't they?*

My brain, too, was doing its best to get back to a new normal.

————

It was July 2015. I had not seen Dr. Volpi in eight years. He looked exactly the same. Just as calm. Just as steady. The same hairline. The same face. The same moustache. I felt like the most changed thing in the room. It felt new to have a fluent conversation with him, to be able to find all my words, to be capable in his presence for once. To no longer be in a hospital gown or my pajamas before him. To him, I must have been a different person.

I nodded. I was there for recent vertigo, unrelated, I hoped, to my stroke.

"You know," he said, "for a long time they didn't think neuroplasticity existed."

No one really understands how exactly healing occurs in the brain, only that it does. The concept of neuroplasticity, of the brain healing itself, is indeed a new one. It was not until the 1970s that the idea that the brain can change became documented fact.

Until the 1970s the brain was understood to be a static organ, fully developed after the first critical years of life, when connections were fixed in place permanently. It's understandable that scientist would think this—there is no other part of the body that continues to change throughout the course of a lifetime with such persistence. Furthermore, before the 1970s most patients who suffered traumatic brain injury died, without modern medicine and life-sustaining technology.

Our bones are our bones, and when they break and heal, they retain their original shape. Our heart is our heart—it too does not morph. If the heart fails, there is no regeneration—we must get a heart transplant. Same with our kidneys, liver, and other organs. We do not grow additional limbs. The notion that the brain regenerates, let alone changes structure in response to learned behaviors and environment, was looked upon as improbable.

Yet my brain, like the MacArthur Maze after a truck exploded and melted the roadways, found alternate pathways around the scar.

I had the fortunate and amazing opportunity of not only experiencing neuroplasticity firsthand but seeing my brain heal in pictures.

In 2015, eight years after my stroke, I had an MRI and I got to see my brain again on scans. When I was pulled out of the bore, when the tech removed the foam from around my head and the plugs from my ears, and I was sitting up with the blanket bunched around my knees, I told the tech I was writing a memoir about my stroke.

She was excited. She said, "I can't even tell you've had a stroke!"

I had heard that line so many times throughout recovery, and it had hurt to hear it. But eight years later I was tickled to hear those same words. "I'm fully recovered," I said, "as recovered as I'll ever be."

"Would you like to see all your past scans? And today's?"

Of course I would. I was surprised and shocked and fascinated. It was eerie seeing my brain on the screen, as if the pictures were of someone else's brain. They looked like black and white Nero Marquina marble—the MRI measures water content in tissues; the fatty areas of the brain do not contain water, and vice versa. In this way the MRI shows contrast between the areas of the brain with fat and those without as white swirls against black. In the first MRI I saw a large white area, like a wispy cloud, in the center of my brain: dead tissue. And a year later, that cloud had become a messy splotch. Years after that, the white was a sharply defined dot. The ischemic penumbra disappeared over time. The infarct became defined.

How could I have a dead spot in my brain and still be the same person I used to be?

"This is remarkable," said the tech. "I wouldn't have noticed your stroke on the scan today. I haven't seen this level of resolution before. It's probably because you're young. I've seen hemorrhagic strokes and they never really go away. That's neuroplasticity."

The brain can change and make up for deficits, sometimes taking over parts of itself meant for other functions, like people who cannot see developing heightened hearing.

Our brains can learn different ways to process information. This is not to say things will be exactly as they were before, but we can regain abilities. When the eye starts seeing things upside down, the brain will eventually adapt, and make it so we see things right side up again. The brain is capable of constantly picking up new abilities and new information and, furthermore, incorporating that knowledge. And when the brain is hurt, it builds new pathways.

The brain changes in response to things like sensory deprivation. In 2008, Lotfi Meralbet, Alvaro Pascual-Leone, and their colleagues discovered that even in sighted adults, as few as five days of being blindfolded changed brains. The participants compensated for their lack of sight by relying more on nonvisual forms of behavior and cognition. And their brains showed these physiological changes.*

The brain atrophies, but it can reorganize itself to make up for deficits. Edward Taub was a psychologist who did research on the Silver Spring monkeys, in what became a historic case in animal research history. In the first police raid in the United States against an animal researcher, the seventeen animals were removed—this case becoming a transitional moment for PETA, expanding it from a group of common-minded friends into a national movement. The group launched the first animal research case to reach the United States Supreme Court.

In Taub's lab in Silver Springs, Maryland, the monkeys were used to study neuroplasticity. In 1981, before they were seized, Taub cut ganglia that supplied sensation from one arm to the brain—the prevailing thought back then being that the

---

* Oliver Sacks, *The Mind's Eye* (New York:Vintage Books, 2010), 207.

monkeys would not be able to use the limbs they could not feel, even if the limbs were functional. He wondered if this was because the monkeys were still able to use the limb they could feel, so he then used arm slings to restrain the good arm in order to train the monkeys to use the limbs they could no longer feel. The monkeys, in this situation, started using the deafferented, or essentially disabled, limbs, despite receiving no sensory input communication.

When PETA lost custody in 1991 and the monkeys were returned to the lab after living with their deficits for years, scientists were allowed to study the monkeys one last time before putting them down. In the subsequent dissection of the monkeys, scientists discovered that significant cortical remapping had occurred. The neurons in the monkeys' differentiated arm maps began to fire when scientists touched their faces. The facial map had taken over the arm map. The brain had reorganized itself.

Norman Doidge, in his book *The Brain That Changes Itself,* describes stroke as a "sudden, calamitous blow. The brain is punched out from within."[*]

My mind and brain were traumatized for some time following that big knockout punch, starting with shock and denial. I tried to do all the things that I thought I could do but could not. I took *Slaughterhouse-Five* with me to the ER, for

---

[*] Norman Doidge, M.D., *The Brain That Changes Itself* (New York: Penguin, 2007), 135.

example. When I got home from the hospital, I tried to make dinner. I tried to bake cookies. I tried to read more books. But I couldn't read anything of merit—even though I did read *People* magazine. Britney Spears shaved her head. Paris Hilton partied. Jessica Biel and Justin Timberlake were dating. Anna Nicole Smith died. So it goes.

I tried to meet up with friends, then melted down afterward, sleeping for days on end.

I parked my car in parking lots and then could not find it afterward. I wandered the parking lot of grocery stores looking for my unique antenna ball, a black and white foam billiard eight ball, hovering over the other cars, or a flash of my car's electric-blue paint.

I was not aware of what I could do or not do. I had to find out, each time. I was doing things the old way out of habit and because I was still not fully aware of what had happened, because my brain had not truly assessed its own damage, and because my mind was also reeling.

I was not upset when I failed at menial tasks. I was too damaged to realize how much I had been hurt, and too damaged to realize what the path of recovery might look like.

My reaction in the early weeks was more "Oh. Well, that didn't happen." As opposed to "Why can't I do this anymore?"

Mostly, I could not remember. I kept forgetting things. My largest deficit was my short-term memory, which was the thing that took longest to return.

In many ways, because of my damaged left thalamus, I was going through a second toddlerhood. The right side of my brain was in a stronger state.

According to researchers, the height of human brain plasticity occurs at or around the age of twenty-six months. At this point in our lives, the right hemisphere has just completed a growth spurt, and the left hemisphere is beginning a spurt of its own.

So at twenty-six months we have a fully developed right brain and a left brain that has not yet caught up. Toddlers, as a result, are complex, right-brained emotional creatures who cannot talk about their experiences.

Even though it was my thalamus that was damaged, I could not convert short-term memory into long-term memory for quite some time. The hippocampus turns our short-term explicit memories into long-term explicit memories for people, places, and things—the memories to which we have conscious access.

This is illustrated in a famous neuroscience memory case—a young man named N.M. who had severe epilepsy. As treatment, the doctors cut out his hippocampus—and while his behavior was normal in the immediate aftermath, they quickly realized that N.M. had no ability to turn his short-term memory into long-term, like some of the patients with anterograde amnesia in the movie *50 First Dates* starring Drew Barrymore and Adam Sandler.

I have to suppose that the thalamus's position next to the hippocampus might have something to do with my deficit. Even though the brain has geographical areas devoted to specific functions, the thalamus's reach is far, touching many other parts of the brain.

This is the way memory develops in the human brain.

Remember implicit memory? The memory system in twenty-six-month-old children, at the height of plasticity, is primarily "procedural" or "implicit" memory—terms used interchangeably by Eric Kandel, a neuropsychiatrist who won the Nobel Prize in 2000 for his research on the physiological basis of memory storage in neurons. Because implicit memory does not generally require conscious recall, memory from this age becomes an unconscious memory, one that is hard to explain.

Implicit memories can also be embedded in our minds from early childhood trauma. They can be triggered when we encounter situations similar to that trauma. They never go away, even though they cannot be recalled, articulated, or explained.

They are the events that make me quail from heights, that make me afraid to sleep alone.

Kandel says the other form of memory is "explicit" or "declarative" memory, which begins developing in the twenty-six-month-old. This explicit memory consciously remembers specific facts, events, and episodes—we use it to describe what we did on vacation and with whom and for how long. Explicit memories can be put into words.

Writing employs both implicit and explicit memory; in the first year of recovery I would feel like I had stories and narratives and ideas to write, but then when I went to the blank page, I could not find the words to express them.

I'd lost my explicit memories, and I'd lost the ability to somehow transfer them into long-term memories—or to be more specific, I'd lost the ability to retrieve any of them, even if I had new memories.

Back then, I was feeling all the feelings but could not

articulate them to the outside world. I felt trapped within myself. I felt like I was under one hundred goose down comforters.

And I also lost literal words. I got them wrong. I still get them wrong when I write—I still switch my homophones. I write *he'll* instead of *heal. New* instead of *knew. Ode* instead of *owed. Fare* instead of *fair.* I wanted to he'll. I new I would he'll. I ode it to myself. The struggle was not fare.

"Aphasia" means loss of speech, but it is not speech as such that is lost but language itself—its expression or its comprehension, in whole or in part.

There are different types of aphasia: expressive aphasia and receptive aphasia. One in which the person cannot find the words to properly or accurately express herself. The other in which a person cannot understand language. I had expressive aphasia, mixing up my words, or making up words, in those first few weeks.

I was intensely isolated and cut off. Some people can write, even if they can't read; this is called alexia sine agraphia. I could write. I had been a highly social person—an extreme extrovert. In those first few hours and days, I eagerly spoke with people, although I realize I must have been speaking mumbo jumbo sometimes. I was an extrovert suddenly deprived of the ability to connect with others, as well as the energy to sustain interaction of any kind. So I became, albeit unintentionally, an introvert.

After my stroke, social interactions exhausted me.

But I kept a memory book. I didn't want to forget. I wanted to at least have a dialogue with myself.

Recognition is based on knowledge, familiarity based on feeling. But neither entails the other. You can feel like you know someone because she is familiar, but you may not know her name.

Years later, having fully healed, I'm aware of the ways I now remember things. That there is someone I know from my MFA program. His name was—his name was what? I try to remember an interaction with him, hoping that the image will give me context for the name. I can recall his face. I can recall the table around which we critiqued manuscripts. I can remember his voice—his name was—aha!—Scott.

Every time I realize I've recalled correctly, I breathe a silent sigh of relief. I do not take that for granted anymore.

I decided to regrow my brain to become a stronger writer. This gave me a sense of purpose in my healing, and a target against which to measure.

My mind told my brain to recognize that writing was an absolute necessity. The brain does not do what isn't necessary. It took me much longer, for instance, to regain my math capabilities than my writing, because I didn't focus on math during my recovery. My husband took over the checkbook. My friends calculated the tip. I used to be able to do math in my head. I couldn't do that anymore. It still takes me longer than before to add and subtract and multiply and divide.

Thanks to plasticity, I found new ways to deal with the old

things—I wrote everything down so I could remember; I let myself feel emotions instead of tamping them down; I began to ask for help.

Meanwhile, I experienced plasticity in more ways than one— while my brain changed, so did my mind.

Culture shock is brain shock—experiences such as immigration are hard on the plastic brain—Doidge says it is "unending, brutal work for the adult brain."* After graduating from Seoul National University, my parents emerged in a South Korean economy that could not support jobs. At that moment the United States was recruiting doctors, scientists, and nurses. My father and mother, an engineer and a nurse, took the leap and then a plane to New York City. This, after surviving a war. They were endlessly working, endlessly adapting to new terrain, learning that "How are ya?" means "How are you doing?"

As literate as they were in Korea, they were suddenly not so much in America. America was not the thing they expected. The crime in Manhattan was as fearful as war itself. As respected as they were in Seoul, they were treated like second-class citizens in America. People told them to go back home. Every day. My mother cobbled together kimchi using savoy cabbage instead of napa cabbage, anchovies instead of salted shrimp. When I taught my mother to swear at people telling her to go back home, she said to the next person who told her

---

* Doidge, *Brain That Changes*, 299.

to go back home, "Go to the hell!" Cultural adaptation and assimilation take huge amounts of brainpower.

———

At one point during my recovery I began seeing my therapist again on a regular basis. Karl had helped me navigate my years as a twentysomething—helping me understand my childhood as it pertained to the emotional terrain of early adulthood and my future. And now, unable to look back or forward, he helped me through the here and now, through what I felt as I was feeling it. It was a relief to return to my familiar psychologist after so many new doctors in my life.

Psychoanalysis is also a neuroplastic therapy.* It works by helping us understand our dreams and losses and thus learn to love and change how we behave. In doing so, we free ourselves of our prior pain and fear and the symptoms of anger and avoidance.

I talked with Karl. I told him about my loss in a way that I could not with friends or my husband, who was dealing with his own loss, the death of his mother and the death of the former me. I began to understand my feelings and grief. In understanding, I accepted my new state of being. I was not the same, and that was okay. It would be okay. *B'seder.*

Freud himself had concepts of plasticity. He had four laws. First, neurons that fire together, wire together (also

———

* Doidge, *Brain That Changes*, 217.

known as Hebb's rule). Second, he connected and related the psychological critical period and the idea of sexual plasticity, or how childhood impacts the ability to love.

His third idea was a view of memory as plastic; the events we experience can leave permanent memory traces in our minds. Memory, according to Freud, is not written down once, or "engraved," to remain unchanged forever, but can be altered and rewritten by subsequent events. Events can take on a new meaning for patients years after they occurred. For instance, when I was twenty-one and graduating from college and making my way into the world, I had to understand what it was my parents raised me to become and why they did so in the way they did and accept that they did their best and resolve my anger and frustration. But when I had my daughter, I found myself reliving my childhood moments yet again—the same ones I thought I'd resolved. I struggled over how my parents exerted so much control over my social life, and how they worried and became angry when I failed to do things as they'd planned— these were all reprocessed as I began life as a mother, as I began to see my own parents in a different way. My newfound perspective also altered my memories of these events.

Freud also believed that people can make unconscious traumatic memories conscious and retranscribe them. When patients had insights into their problems, they began to regard him as they had important people in their past, usually their parents. The patients were reliving memories without being aware of doing so. They were transferring scenes and ways of perceiving from the past onto the present—"reliving" instead of "remembering."

Sometimes my diminished capacity was too new and sudden for people in my life, such as Adam—but it was the way it had to be. It brought on plasticity for others, too.

Adam had lost his wife to brain trauma.

When I had a baby, Adam lost his wife again, to motherhood.

My stroke brought me the experience of neuroplasticity—of my brain adapating and changing as it healed—but also the experience of plasticity: of having to change and be malleable and form a new self and a new life. That I could restructure from the ground up and become a new version of myself, the same way my brain did, by rerouting and retooling and recruiting other structures to take over the duties that were lost in damage, was a new insight. My stroke taught me resilience—unbending emotional strength has its place in life, but plasticity has even greater value.

By six months into my recovery, I was able to read a short story. On my MRI scans from June 2007, the scars on my brain were still very visible, albeit reduced in size by almost half.

At eighteen months, I felt like I'd regained 80 percent of my pre-stroke abilities. By then I could read a novel. I could write a basic short story. The MRI scans illustrate the progression—the white cloudy scars were diminished at this point, the ischemic penumbra regenerated, leaving sharp edges to the black scar, as opposed to a white and gray foggy cloud.

Seeing the MRI scans in sequence, was stunning. They illustrated my comeback in a way nothing else could.

I see my daughter grow up—from someone who I was like in my first weeks of recovery—to someone who at the age of three narrates her inner thoughts, which seem unrelated in time or subject. It is her prefrontal cortex, which continues to develop until the mid-twenties. Red yes turn here and Mama and I want candy and music is happy. It sounds insane, but it is only what goes through our own minds, leaping back and forth in time.

———

It is December 31, 2006. I have a funny feeling in my head. It is a stroke. It changes my life.

It is 2013. I give birth to my daughter. What follows is a dark and persistent depression from which I cannot emerge, no matter how much I exercise or dance with my infant in the mornings or count my blessings. The blessings fade. What stays with me is exhaustion and hopelessness and self-doubt, the kind I haven't experienced since my stroke recovery. All I want to do is sleep and stay in bed. I do not believe things could possibly get better. Until they do.

It is 2014. I am separated from Adam. From the woodwork come my friends, some very unexpected. They hold me when I cry. Two hold me while I go to sleep. One writes my daughter a poem.

It is 2006. I am at Hedgebrook at a writing residency, with a headache I cannot shake. In hindsight, I am not sure if this is a precursor to my stroke later that year. It is at Hedgebrook that I feel a deep loneliness and void within me. The residency becomes a vision quest.

It is 2008. Stroke recovery is slow. I wonder if I can hold a thought in my head, if I can imagine worlds and then write them down.

It is 2007. I learn that I had my stroke in a key area, the thalamus; more specifically, the dorsomedial thalamic nucleus.

It is 2013. My father has a stroke. He spends a month rehabilitating at an inpatient facility. He cannot speak. He cannot swallow water. He will learn to speak again. He will drink out of straws for over a year. He will need a cane, likely for the rest of his life.

It is 2015. One of the friends who held me while I slept dies from a drug overdose. He was forty years old. The friend who wrote my daughter a poem dies from a massive stroke. He was forty-six. My postpartum depression, persistent even with medication, finally lifts. I open my eyes and there is my daughter, whose eyes open even wider at my return. This is the mother she has not yet known.

My brain showed me resilience in its plasticity.

A decade later, when I was going through a traumatic life transition, a close friend who I will call Lisa told me, "Christine, you have true grit."

I told her I'd never received a better compliment in my life. And I thanked my brain for showing me the way.

10

I began to dream, or remember my dreams, that summer, six months after my stroke. What I could not remember in the day, I could remember in sleep. I dreamt about Israel and the Mediterranean Sea. That I lost a red scarf out my hotel window and it floated out over the sea until it was a speck.

In real life, I wrote my mother-in-law a letter. I buried it by the side of the road in late October of 2007, one hundred feet above the embankment where she died. According to Jewish law, one must pray for a year so that the soul can make its way out of this life. According to Jewish philosophy, we assume our loved ones are not so evil as to require an entire year of prayer. Thus, we actually pray for ten months. She was still here. It had only been a few months since her death. I did not want to pray anymore. I wanted a direct line to her, to say my good-bye, to apologize, to make amends, to say, *B'seder. Everything is in order. It will be okay.* There are so many ways to say "okay." There is the sarcastic "Ooookay." There is the enthusiastic "Okay!" There is the annoyed *"Okay!!"* There is the skeptical "Okaaay." There is the quizzical "Okay?" There are, likewise, so many meanings to *"b'seder."* It is okay. Everything will be okay. Everything is fine. Fine. Okay. Okay. Okay. Okay? It means everything is in order, but it is used in so many circumstances where everything is not okay. When someone has died, for example. How many times did people say *"B'seder"* to me in Israel? I cannot count. *B'seder. B'seder.* My life. Her life. Our lives. *B'seder.* In order. Okay.

A few weeks later a devastating fire ripped through Topanga Canyon and Malibu. The letter burned and turned to ashes, floated into the sky. I dreamt she read the letter, and that it came to her in pieces, like a serial novel.

I dreamt about being pregnant with twins. That I gave birth, and that they began talking immediately. They spoke to me in their own language of made-up words, and I understood what they said with fluency. No one else could understand us and we were all put in an institution, but we did not care. We had one another.

It was as if my brain, after dormancy, was now producing a lush, imaginative world in my subconscious. I wrote these dreams down immediately upon wakening. I did not ever want to forget them. They inspired me. They told me my brain was reawakening. While unconscious I reveled in these newfound images, but I struggled in my waking hours. I could not write. I could not find the right words. I was still ordering only hamburgers off of restaurant menus. I still could not add. While the new me was struggling, the old me, what was left of her—was judgmental and angry and coarse.

I did not deserve to live. I did not want this new life. I did not want this new me. I missed my old brain. I could not do anything on my own anymore. I raged.

This was when I began to see my old therapist again.

"I'm depressed," I said. "I've lost all my support columns," I said. And then I cried.

My therapist saw me cry more in those few months during stroke recovery than he had in all my previous years with him.

"I want to die," I said. "But I don't want to die. I don't want to feel this way."

I'd wanted to die in my early twenties. I went to Karl because he was like no other male figure in my life. And I wanted a male mentor I could learn to trust. He was gentle and he asked about my feelings. He was Jewish and understood my Jewish life. He played the banjo and was a socialist and I did not understand those things, but I loved that difference, too.

I would crack jokes between horrible confessions. I can't write, I would say, but boy oh boy, Paula Deen has become my best friend. I don't know how much time has passed, I would say, but, well, I guess I'm unable to be bored. Because that is what I do. In my family, all was forgiven if you could make something funny. If you got into a fight at school, that was bad. But if what you did was bite the other kid on the belly button, and it made my father laugh? You were forgiven. I was fat. But if I said to my father, "Is everything else about me perfect?" And he said yes, and I said, "Then G*d had to make everything fair," my father laughed. I was forgiven.

"I'm delighted with your humor, Christine," said Karl, "but you don't have to entertain me."

Was I entertaining him? "Am I? Entertaining you?"

Indeed, I was.

"How do you feel?"

I did not know how I felt.

And now I was back in his office. Talking about my feelings. Building a column of support. My life had fallen apart, and I was, unbeknownst to me, building it back up in a brand-new way. There was so much plasticity in my life. My psyche was plastic, too.

"I've lost access to my toolbox and coping mechanisms," I said. "Except for my humor. Haha." My humor, it turns out, was the key to my resilience.

"That's what the thalamus does," said Karl. "It helps with soothing."

"I'm in free fall. I can't figure out how to deal."

And so Karl helped me understand what was going on in my brain. And over time, I built a new toolbox. I put in new tools.

My regular coping mechanisms were inaccessible, and understanding what was happening on a cognitive level helped me accept myself and treat myself with more tenderness. Without my previous tendencies to build a wall and compartmentalize, I could find a new way to process grief, to actually experience myself in the present tense.

This was an opportunity for change. Change was okay. Things did not have to be black or white; things were often gray. The gray of not knowing whether or not I could regain my abilities had lessons and value in and of itself. This was a time to get to know myself again. This was a way to explore.

Plasticity, at work.

Meanwhile, I was walking on eggshells with regard to my physicality. My doctors instructed me not to take my heart rate above 120 beats per minute for fear of causing another clot to go up in my brain. For a thirty-three-year-old, 120 beats per minute is 64 percent of maximum heart rate. So I literally had to confine myself to walking. I didn't speed walk, and I didn't walk up hills, and I definitely could not run. I was not to strain myself. Despite getting better, it was frustrating to not be able to move around, to be reminded of such strict physical boundaries. And my depression didn't improve as rapidly, because I could not exercise. And because I could not exercise, I got out of shape. And because I got out of shape, I beat myself up.

And still I threw another clot up into my brain.

On June 2, 2007, I awoke and could not see out of my right eye. Things were blurry, and my field of vision had narrowed. I thought maybe there was something wrong with my contact lens. I shrugged.

There was not the mystery like I'd had months earlier when I first had a stroke. I noticed that something was wrong. But again I thought maybe it would go away. I couldn't possibly be having another stroke. It was probably my contact lens. It was probably exhaustion. And so I ignored it. I did not tell Adam. I just wanted it to go away. Old habits die hard. Denial is not a river in Egypt. And the problem did go away.

The next day I could see out of my right eye again. It was over, I thought. Whatever that was. I shrugged.

The next morning after that, I couldn't see out of my left

eye in exactly the same way I'd experienced with my right eye. My vision was blurry, and I'd lost some of my peripheral vision. I could not see that Adam had already left for work until I turned my head and saw that his side of the bed was empty.

I took out my contact lens and put a new one in. No difference.

I took out both my contact lenses and put my glasses on. No change.

I couldn't see very well. I wouldn't be able to make out freeway signs with my left eye like this.

When I called Adam at work to finally tell him what was going on, he said, "I don't know how this is not an emergency room situation."

Oh. An alarm rung inside me—a new alarm that brought me into action, that told me I needed to take care of myself. It felt strange, actually, to pick up the phone. It felt completely self-indulgent to ask for medical attention. I was still so afraid of showing vulnerability. To say "Something is wrong" was to say something in my body was failing was to say I needed someone was to say I deserved help. I did not think I deserved to be saved.

I have thought about what it is that makes me ignore warning signs in my body. Was it a lifetime of living with invisible physical deficits that no one else had or noticed—and thus that I too overlooked? Was it being raised by parents who taught me to be so strong that I ignored all pain? Was it prioritizing everyone else over myself? Was it my own self-destruction? Was it a combination of all of the above?

But in the end I wanted to save myself. I had to, I realized, call the doctors or die. It had come to that. I called every doctor on my roster—my primary care physician, my neurologist, and my ophthalmologist.

At eight thirty in the morning, none of my doctors' offices were open. I left messages. My ophthalmologist was the first to return my call.

"Come in," he said. "Right now."

He unlocked the door of his office, when I arrived ten minutes later, alone. My ophthalmologist Dr. Yokoi, had apparently driven into the office early to meet me. And yes, the left eye was indeed weaker than at my last visit, just a few months prior. By -1.5 points. He said this could be diabetes if it were both eyes.

"But it's only one eye. This is odd. Call your doctor." He suspected blood sugar issues. "Head to the emergency room."

My neurologist called me as I headed back home to wait for Adam to come home and drive me to the hospital. Fifteen minutes had passed.

"Meet me at the ER," he said. "Sounds like you threw a clot up into your eye—get yourself to an ER immediately, where you should get an MRI." He told me to go to the hospital at which my cardiologist worked.

So we went. The hospital where my cardiologist worked had my least favorite ER, the one that had made me wait hours last time, the one I ditched for John Muir, where I was first diagnosed with stroke. Now there was a pregnant lady in the waiting room, who was bleeding, who was kept waiting, and was still waiting an hour later.

I waited an hour before I told the ER, "My neurologist sent me here. They think I'm having another stroke."

Meanwhile, Adam walked two floors up to my cardiologist's office to tell her I was in the ER downstairs. Her office told him, "She's not on call. She's in a continuing education class. She's not coming down."

They never sent another cardiologist from the group down. I never saw her again.

I was told I would be called shortly. An overwhelmed nurse told me, "There's no way you'll get an MRI today, the doctors say. Maybe not tomorrow, either."

We walked out. And we drove thirty minutes to John Muir, again.

Relief flooded over me the minute I walked through the doors. It was familiar. There was no wait this time, either. Within five minutes, three nurses converged upon me, took my vitals, and recorded my health history. Immediately after that an ER hospitalist came to see me. A few minutes after that my neurologist walked in.

"Please," said Dr. Volpi when I saw him in the ER, "you're not bothering me. If you experience any symptoms out of the ordinary, call me."

"I don't like to complain."

"You're not complaining."

I was wheeled in for a CAT scan. An hour after that I was in the MRI machine.

And then I was admitted and wheeled up to the DCU. The nurses greeted me. "Forty-Seven!"

"Hi!" I said. "I'm back!"

That night my vision returned to near clarity.

"I can see again!" The room was dark. I announced this to no one in particular.

I stayed for two days, during which time I learned that I had thrown another clot into my brain and had a TIA—a transient ischemic attack, sometimes referred to as a mini-stroke. The clot traveled from one eye to the other, from one optic nerve to the other.

I was on a conservative treatment regimen of blood thinners—but this TIA made the need for a PFO closure urgent. They needed to close the hole in my heart, because it was an imminent danger. It was allowing clots to seep through.

A transient ischemic attack is caused by a clot—the only difference between a stroke and a TIA being that the loss of blood flow is temporary in a TIA. The symptoms occur rapidly and over a short period. Most TIAs last no longer than five minutes, and symptoms resolve within twenty-four hours. TIAs do not cause tissue death, and so they cause no permanent injury to the brain. But because they have the same underlying cause as strokes, they are called mini-strokes.

My TIA caused amaurosis fugax, a sudden transient dimming or loss of vision in one eye.

Part of why I was not as alarmed as I should have been is that TIAs had been regular occurrences in my life. This was not the first time a clot traveled up into my brain. Weird things had happened before. Weird things that I figured would happen again. My brain would hurt. My fingers might tingle. I might lose my words for an hour. So it goes.

In the hospital, after this diagnosis, I traveled in time again. The present was so much like the past.

───────

It is 1995. I am twenty-two years old, on a bicycle ride with Adam. I have just celebrated my birthday. It is our third date, and he is on the Cal Cycling team. I keep up with him throughout the ride. We park our bikes at Berkeley Bowl. And when I get off the bike, I cannot see, and I cannot feel my arm, and I have no sense of balance.

I went to the doctor then. I made an appointment with a neurologist. At Stanford they gave me an MRI. When I called to ask what had happened, my neurologist there simply said, "Well. We have ruled out meningitis."

"What?" I asked over the phone. "You thought I had meningitis? Oh shit!" And that was that.

My guess now is that I'd had a TIA, which by definition left no permanent trace, no way to be diagnosed.

───────

But now I was diagnosed. I was in the hospital again, with Dr. Volpi. Adam was there again, only we were older, married. I was in a hospital gown. I was playing with the oxygen mask.

I had become smaller and smaller as the years wore on, as we focused on Adam's career. One of the first written correspondences I'd had with Adam revolved around my philosophy of love and relationships; I said I wanted someone I could take care of, and someone who would take care of me in return. To be enamored of someone completely and ignore myself, because that someone would be enamored of me. I was twenty-one years old when I shared this, and I was mistaken.

I relied on Adam to take care of me, because I would not take care of myself. This dependency caused immeasurable strain for years to come. To rely on someone to take care of me meant to have my needs defined by someone else, that I didn't prioritize my needs, that my partner had to anticipate my needs. It meant exhaustion, meant resentment.

It was something that had to change.

That change took a long time.

Leonardo Botallo, an Italian surgeon, is credited with the discovery of the foramen ovale in 1564. He noted a communication between the left and right atria, formed by an overlap of skin between the septum primum and the septum secundum.

In 1877, Julius Cohnheim, a German pathologist, described a case of a fatal cerebral embolism that he believed had passed through a patent foramen ovale, or PFO. He was the first to propose that a venous thromboembolism could bypass the lungs through a PFO and enter the arteries.

In the late 1980s, scientists conducted a case control study that demonstrated an increased prevalence of PFO in stroke patients without an obvious cause such as carotid stenosis or small vessel disease. This suggested that PFO was an associated cause of stroke, especially in younger patients.

My patent foramen ovale had to be closed. I had thrown another clot into my brain, despite blood thinners and a prescribed sedentary life.

My neurologist referred me to an interventional cardiologist, Dr. Neal White, who scheduled a PFO closure within the week. He was, it turns out, the cardiologist leading the way with this procedure in the Bay Area at the time. He was, it turns out, mentoring my prior cardiologist with this procedure. My prior cardiologist, it turns out, had delayed my procedure because she was trying to schedule a surgery time where she could operate alongside a supervisor—alongside him, specifically. Needless to say, after her no-show in the ER, I wanted Dr. White to be my doctor.

In 2007, PFO closures were a relatively new procedure in the United States. In 2016, when I told my new primary care physician that I'd had a PFO closure, the first remark out of his mouth was, "You must have been one of the first people to have their PFO closed that way." I was.

Implanted inside my heart was a device called an Amplatzer septal occluder. It looks like a double-sided umbrella, without a handle. Or maybe like a butterfly. The "umbrellas" ultimately straddle the hole, over which scar tissue grows.

Until 2006, only two transcatheter devices had received limited FDA approval for marketing to close PFOs. They approved the Amplatzer, along with the more commonly used CardioSeal septal occlusion system, for use in the early 2000s. The limitation involved an HDE (humanitarian device exemption), which is a category of FDA approval that applies to devices designed to treat a patient population of fewer than four thousand per year.

It was used well in excess of four thousand patients per year.

As a result, in 2006 the FDA withdrew the HDE approval for the device. As of 2006, closure devices have had only off-label uses, even though the FDA says there is no problem with doing so. There are no PFO closure devices currently approved by the FDA specifically for PFO closure, even in 2015. Devices approved to close other types of holes, such as atrial septal defect or ventricular septal defect, are used to close PFOs.

Before my PFO closure, my cardiologist made sure to measure and examine the hole with ultrasound. He redid the tests to be sure. Another transesophageal echocardiogram. Another ultrasound.

In the hospital on closure day, Dr. White wove a catheter, a long, thin, flexible hollow tube, into my heart. The catheter is inserted into a large vein through a small incision made in my upper inner thigh in the groin area, precisely where my thigh meets my torso. My cardiologist, I thought, was going to see a very intimate part of my body. He might see my vagina, even. I thought this as the anesthesia took effect.

The PFO closure device is moved through the catheter to the heart and the location of the defect. Once correctly located, the device is set in place so that it straddles each side of the heart. The catheter is then removed and the incision bandaged.

The Amplatzer device is made of a nickel-titanium alloy. It can go through MRIs and X-ray devices. It's a slightly bulkier device than other popular models cleared to close PFO in off-label use. It is more popular in Europe than in the United

States. It will stay in my heart for the rest of my life. For the first year or two afterward, I needed to take antibiotics before certain procedures, such as teeth cleaning, in case of infection.

The device remains in the heart, where within a few days the body's own tissue begins to grow over the device. Over the next three to six months, scar tissue will build around the device and then the hole is closed forever. Until then, I was not allowed to engage in contact sports.

"No problem," I said, "I don't play football or kickbox."

———

In the catheter lab I was awake but sedated. I do not remember a thing, but I had become used to that. In the midst of the procedure, my cardiologist had to defibrillate me. My heart fluttered, or had atrial fibrillation. In other words, my heart was tickled by the implant and freaked out. I know this because my cardiologist told me.

"Do you remember it at all?" he asked.

"No," I said. "The Versed did its job." But maybe, I joked, I wouldn't remember it anyway.

"Good," he said. "You were talking to me the whole time."

"Oh my G*d, what did I say?" I asked.

"Don't worry about it," he said.

Not a reassuring answer. I wondered if I'd betrayed any secrets. But when I stopped to consider it, I couldn't even remember my secrets.

I stayed overnight in the hospital, flat on my back. I was not to move. Not until the femoral vein closed. But I bled anyway. The nurses checked me every hour and found me soaked in blood in the middle of the night, the incision in my groin having opened.

I zoomed back in time to my teenage years. I was in the hospital. It was 1990. I'd just had a tonsillectomy. The United States had just invaded Kuwait. I know this because the news was showing images of fire and bombs and military tanks when I woke up in my hospital room. Unbeknownst to me, one of my closest friends was driving through the desert escaping Kuwait for a new life in Egypt and then Connecticut. R and I would meet twenty years later at Hedgebrook on an island in Puget Sound.

My mother was by my side when I woke up from my tonsillectomy.

I was nauseous, I said. I needed to throw up. My mother handed me a kidney-bean-shaped bowl, into which I vomited.

My throat burned, as if someone had stuck a hand down it and scraped it with talons. My mouth tasted metallic. What is it that tasted like metal? Blood. I looked down. The bowl was filled with blood.

I started to cry. Why was I throwing up blood?

My mother handed me a numbing throat spray.

She rubbed my back.

I threw up again. And again. All the blood that had collected in my stomach from my tonsillectomy was coming up. There was blood everywhere. On the sheets. On my gown. On my chin. I wanted to scream, but my throat hurt too much.

And then I lay back. I was sleepy. I went to sleep in the hospital room, dimly lit by a fluorescent light that never turned all the way off.

I woke up in the hospital again in 2007. The nurse was pressing down on my leg.

"Have you moved?" she asked.

"No," I said. I had been sleeping. I felt a sickening pressure on my leg.

"Hold still," she said. She exerted pressure on the seam where my leg joins my torso. The nurses took turns doing so over the next hour or so. They were on their tippy-toes, their bodies bent over mine, pushing all their weight against my crotch. Until I clotted. I went to sleep in the dim light.

Six years later, in 2013, when the labor and delivery nurse pressed on my stomach after I gave birth to my daughter, I would be reminded of this very same act. She was pressing with all her might, to help my uterus contract back to normal size.

"I'm sorry," she said, "I have to do this."

"It feels good."

"Really? Most women hate this part."

"It feels good. Press harder. You're saving my life."

The PFO closure was critical in my healing. Incidentally, it was also how I never got a migraine again. PFO closures are sometimes done to prevent migraines, and it often works, even though doctors don't know exactly why.

My PFO was the reason that for years I had not been able to exercise to exertion. That I would get migraine headaches. That running felt like drowning. My blood had not ever been fully oxygenated.

I started running a year later, because I could. Because every breath no longer felt like drowning, like a gasp for air while being held down by the throat. I was amazed at how easy running could be now. I could understand how it was that humans could run marathons, whereas before I thought it was a superhuman feat. I retraced the backpacking trails I'd struggled up, jogging them now. Lifting heavy objects no longer left me light-headed to the point of nausea and fainting. This was how the stroke saved my life. This was how the stroke changed my heart. My heart now had no hole.

My brain was changing, but now so was my body.

11

I am a child of immigrants who raised me to be tough and impervious to pain. This mind-set informed my reaction to my own stroke and my recovery.

When I was a young child, my parents took my brother and me on weekend hikes in the San Gabriel Mountains. Mount Wilson loomed large, just five miles away, like a wall above our home. There were days you could see the silhouette of the mountains, when you could see the oak trees dotting the mountainside and the television tower like a birthday candle atop its peak. Then there were days when the Southern California smog obscured the mountains with a pinky-brown haze that made it hard to breathe, that for me set off a sharp stab with each inhale. On clear days I could also see the fire roads and switchbacks zigzagging up to the tower from the San Gabriel Valley—and it was on these trails that we hiked, four specks on the mountainside, dust and granite lining the way, the sun always merciless, the air thin, and my lungs burning.

My parents enjoyed the walk, and my brother bounded up on light bunny feet. To me, it was suffering. But I did it, because it was supposed to make me healthy, and because I hoped it would get easier and easier with every hike. This hike was something I was supposed to be able to do without complaint, given that older people and children treated it like a stroll. The thing was, it never did get easier. Not after weeks, not after months, not after years. I panted each time as if it were the first. I vomited along the trail. Sweat poured down my face. I rested and then scurried to keep up, because I did not want to be left behind. I tried so hard to keep walking, and I did. I did, because everyone else could.

"Why can't you hike? Why are you so out of shape?" My parents asked those questions. And then I asked them of myself.

There was something wrong with me, I thought, that I could feel pain so easily, that I tired so easily. I thought the something was weakness, some shortcoming in my mind and body. I was lazy. I was not good enough. Everybody else had a capability I did not. I tried harder.

"When we were your age," said my parents, "we had to walk like this all the time. During the war, people walked farther than this to get to Seoul. To escape from the North. People got separated from each other because they couldn't keep up."

I would stare at my birthmark then. It's about two inches in diameter, the size of a watch face, on my left wrist, where a watch might go. "This," my father once told my brother, "this birthmark. Remember it. This is how you will know who your sister is, if you become separated by war."

Decades later I would repeat this anecdote to a friend of mine, someone whose own family had been in the Japanese internment camps. I anticipated a solemn nod or other form of empathy.

Instead, she burst into giggles.

"What?" I asked.

"Oh"— she gasped for air—"seriously? That's the funniest thing I've ever heard! That is so extreme! The war is so over!"

And then I too began to laugh. The war was over.

My own parents lost siblings during the Korean War— there are still family members whose fates we do not know. They suffered decades of separation. This loss is one of the biggest tragedies of war, and my father was preparing us for an event, however unlikely, for which he had been unready.

And so my brother and I were raised to survive wars. It was the thesis statement of our upbringing. People who couldn't walk, who sat down and cried—they died. Furthermore, we were raised to believe that war could happen at any time and that we had to be prepared. After the planes flew into the Twin Towers on 9/11, one of the first phone calls I received was from my father. "Go to the store and buy water. There's going to be a war!" Not probably. But definitely. And so we hiked.

No matter how much I gasped for air, panting as if I'd been on a long run, I couldn't breathe. I was suffocating very slowly in the chaparral. The sun bore down, and the dry dust left a film on my tennis shoes.

I made up stories to push my body forward. I imagined wartime, and fleeing danger. I imagined warriors pursuing me with weapons. I imagined bombs behind me. Planes. I was being chased. There were men with axes pursuing me. Murderers. Men on horseback. I imagined my ancestors walking, bent forward with possessions on their back, and I thought myself light in comparison. I walked in bold terror, not courage.

Meanwhile, my mother would take detours off the trail to investigate wild greens. These days this is a trendy thing called foraging. In my mother's childhood it was survival. In my childhood it was embarrassing.

"Look!" She would brandish a handful of roots or leaves or stems. "This stuff is tasty!"

My father would shake his head and roll his eyes. But later he'd eat all the sautéed greens without hesitation.

I always made it to the end of the trail, which culminated in a pool at the foot of a waterfall. I would put my feet into the icy water and revel in the shade. I would catch my breath, lie still, and my sweat would cool me in the breezes of the shady dark. It was worth it, I would think. It was worth the pain for a measure of peace.

One time we saw a trout there, and my father picked it out of the stream with his bare hands. I could hear it flapping around in my father's backpack, gasping for breath and dying as we scuffled downhill back to the trailhead. We had the trout for dinner. It was delicious. From this I learned that the destination and the goals were worth the suffering of the process—the end was the reward. That determination would see me through to that end. I became unafraid of pain.

Years later as I was walking on a trail, a man chain-smoking cigarettes passed by at a rapid clip. I was in college by then, studying abroad in Seoul. In Korea I'd discovered that switchbacks were very few—the trails there went straight up the mountainsides, sometimes supplemented with man-made stairs. As a result, the trail was shorter, albeit steeper. The air was damp in monsoon season, and I'd become used to the sweat. I didn't smoke. By that point in my life, I'd long accepted that I was unable to walk or run as fast, and as long, as anyone else. And that it was completely normal for a chain-smoking man to scurry past me on a steep uphill.

I had given up on caring for my body by then. I was in my early twenties, and I'd spent more than a decade at war with my body. My mind was a tyrant over my body, and I did not care to listen to my heart or my arms or legs or lungs anymore. I did not care for their needs. I had not cared for my body in nearly two decades.

I abused my body by starving it. I was a fat child as a result of my inability to exert myself through exercise. I asked the skinniest girl in sixth grade how on earth she stayed so thin. Her name was Leanne, and she said, "I don't eat breakfast, and I skip lunch, and I barely eat dinner."

I was aghast. But she seemed so certain. And she was so thin.

It took three more years to take her advice, but take it, I did. I put my mind to work and ignored my body's cries for food. I followed Leanne's regimen in junior high. Each time my stomach growled, I denied it nourishment, until it was tied up in knots, until it gave up and let out the dullest bleats of pain. I was down to eight hundred calories per day, consuming much of it at dinnertime. Lunch was two slices of low-calorie bread, one thirty-five-calorie slice of Louis Rich turkey, one eighty-calorie slice of cheese, a thin swipe of light Miracle Whip, and a squirt of French's mustard. Two hundred calories for lunch. It did not dawn on me that I was consuming an amount of calories my ancestors probably clamored to eat.

My parents were happy to see me losing weight. I bet my father I could lose twenty pounds in eight weeks. He offered me two hundred dollars to do so. He was joking at first. I lost thirty pounds within two months. I weighed ninety pounds. I claimed my money. He said, "You finally look perfect." He was not joking. It was the first compliment from him in years. I missed this acceptance. He and I had been very close throughout my childhood, playing hide-and-seek and blackjack and poker, the latter two for the purpose of learning strategy, he said. But he loved that his seven-year-old daughter could play blackjack. I missed this closeness and approval, which disappeared when I hit junior high school, when my father steered me toward school and achievement and away from carefree fun. Because survival. Because war. Because success.

And all that while, I was getting headaches. I had my first migraine at the age of thirteen; auras, nausea, and then debilitating pain on one side of my head. They would bring me to my knees, and then I would have to take to bed. My ophthalmologist diagnosed them as migraines. I was drinking Crystal Light, and the aspartame sweetener increased the frequency of the migraines. I couldn't look at cars flashing by in sunlight, because it would induce a migraine. I couldn't ride a bicycle up a hill, because I would get a migraine. My parents' and my worst fear was that I would get a migraine before a final and not be able to function for test taking.

My parents were worried that I wouldn't fit in. They were worried that I'd lose the war for acceptance, just as they had as immigrants in 1960s America. They were worried that I would end up alone and starving and die. And so they pushed me to excel in school, to be well coiffed, to be polite, and to be thin. They wanted me to be safe—and safety to them resided in the middle of the pack with regard to appearance and the top of the pack in intelligence and achievements. When we watched *Mutual of Omaha's Wild Kingdom* on weekends together, my dad would point to an antelope or gazelle and say, "See? The one at the side or the back gets eaten. Be at the front and middle."

So I tried hard to fit my body into the most acceptable shape. One that would not stick out. In other words, I tried to be thin. I wore fashionable clothing that was not at all unique. I wore high-top Reeboks like everyone else. Double socks like everyone else. Leggings. I did not wear a hat, like the one girl who wore hats at school. I did not wear goth clothing—even though rebellion was so tempting.

When I was a toddler, my parents carried me to a neighborhood preschool in Queens and asked the director, "How can we make sure our daughter doesn't have an accent like us?" So as not to stick out. I imagine them now, trying to think of all the ways they could protect me from harm. And protect me from the things that harmed them.

The solution was that my parents would not teach me a word of English. The school would teach me. And that is how I was born in the United States but did not learn English until I was four years old.

This is how I learned to speak English: when I got to school, I only knew the phrase "Where is the bathroom?"

I did not know how to say anything else. And so of course I had a tantrum at school, because I was frustrated that the teachers could not understand me, after having lived with my family who could understand every word I'd said. I wanted to say that someone needed to be called, and I remember saying, *"Bulow bulow!" Bulow* is the word for "call someone" in Korean. And so they locked me in the very room for which I knew the name. More specifically, they locked me in a bathroom stall. And they didn't let me out until I calmed down. I spent the first three days of preschool on a toilet. Whenever they came in to check in on me, I would kick the bathroom door, pound the walls, and scream. I learned from that experience the need to communicate. That no one would listen to me until I spoke their language. That I had to speak it with perfection. That words were the way out.

And when I had my stroke, I lost my words. And yet, words were how I measured my progress; words lit the road before me. I wrote to save my life. Despite my parents' best effort to protect me from harm, I was harmed in an unforeseeable way. I learned that I had to protect myself.

My parents pushed me to excel in school—in America, they said, education was the key to upward mobility. Material possessions and money, they said, can be taken away from you, but your education stays with you. My high school was competitive. Amid the formulas to get into college, I stifled my whole being to fit into the perfect persona. To get into every single college. Because my goal was to get out of that miserable town. By grade school, we'd moved to Arcadia, where

I forever missed New York City. Arcadia was a safe town, the perfect place for middle-aged people who had survived a war and immigration and wanted nothing exciting to happen again, ever. But I was a teenager. And safety was not my priority. I suffocated. I thought if I could just leave Arcadia, I could breathe. I could be rid of the smog, and I could have my own rules, and I could be free.

In that climate—where I never felt good enough—I wanted to cry, and sometimes I did. My mother would hold me, and then she would put her hands on my shoulders, look me in the eyes, and give me advice in a gentle voice.

"It is okay to cry," she would start. "But the world doesn't want to see you this way. So when you want to cry, take a pillow. Then take the pillow into an empty room. Lock the door. Then put your face into the pillow and cry so no one can hear you. Also, write this all down. This is good material for your writing."

I cried so no one could hear me. I choked down tears until I couldn't breathe. My mind was now a tyrant against my heart. I built a wall. It worked. And then the wall locked me in. Within those walls I battled my body. My mind made it all my body's fault.

I punished my body. My mind was dying under the pressure. The wall I'd built was so thick, and I felt I had to be perfect, and I was so angry, and I'd tamped down my feelings so I no longer knew how I felt. My emotions became a little flawless diamond.

Parents pass on their fears to their children. The fears I inherited had no face, because my parents did not explain to me the cause of their rigid expectations. What was it they had suffered and now ran so hard to avoid? They were protecting me from their own harm and trauma and what they had seen in the wake of the Korean War. And yet I adopted the same frantic dread. And my mind made up monsters to explain my terror. I accepted these monsters as reality. And this spread throughout the other venues of my life. To friendship. To love. To school and academics. Nowhere was safe. Everything was treacherous. Everything had to be controlled. Safety was the edge of a knife.

A terror of failure and lack of acceptance embodied the anxiety I felt around me in my home, which followed me to school. That if I failed, if I was an outcast, if I was alone, I'd be doomed. That the teacher I respected was also a dangerous creature who gave out grades and thus dictated my future.

I worked on my high school newspaper. Yes, our high school mascot was "the Apaches." And yes, the high school newspaper is called *The Apache Pow Wow*. Yes, that is the climate in which I went to school as a Korean American. Yes, that is still the name of the high school's newspaper.

I was the sports editor. I did not like sports, but no one else wanted to be the sports editor. I also wrote features, one of which entailed interviewing alums about how they now—the year was 1990—felt about Arcadia High School. I remember interviewing one alum via phone.

"What was Arcadia High like in 1970? What's different now? How do you feel about it?"

She replied, "Back then, we didn't have so many Asians. We had more school spirit. People went to games and cheered. The new Asians don't care about our school. They just don't get it."

I realized this glimpse was what my perfect English, which stood out amidst all the new Asian American immigrants in 1990, brought me. I paused. "Ms. L., do you realize you're speaking to an Asian American?"

Her voice went up a pitch. "But you're not like the rest of them! You're like the Asians we had at school back then. They were Japanese American, and they were assimilated and could speak perfect English. I couldn't even tell you were Asian on the phone. You don't have an accent. You're like the Japanese I went to school with. You get it."

I sighed.

She heard the sigh. "You won't print what I just said, will you?"

My friend Michele, the editor in chief, urged me to write the real story.

Here is the girl I used to be. I did not print what Ms. L. had said. I edited it down to simply "Back then we used to have more school spirit." Because I did not want to offend. Because I did not want to ruffle feathers. Because I did not want to be noticed. Because I wanted to survive.

Today I regret that I did not print the words as they were spoken to me. My high school was rife with racial tension. I could have helped school community and helped myself by saying the truth. And today I do.

By the time I got to college, I had come to depend only on my mind and brain, the things that had never let me down. I ignored my heart—and I intended to be a premed student. I was going to use my mind. I was going to enter a field in which emotional compartmentalization would be of use. I would be of use.

At Berkeley, I ended up majoring in English.

I never did end up applying to medical school.

I developed a wicked sense of humor alongside this sense of control. In my family, if something was funny, it could be forgiven. Laughter was an immense emotional outlet and a survival mechanism.

My brother and I were supposed to, in theory, take turns naming our pet dogs. But every time we got a new dog, my brother would propose standard names, and I would offer something more outrageous. My brother never got to name a single dog (sorry, little brother). Though my father had promised that we would alternate names, in the end the funniest name would win. Of course, our dogs had Korean names that translated to Fart and Dude, instead of Rover or Dog.

Knowing my mother had started work as a nurse in a county prison, and knowing she would not talk about her work there, I of course as a teenager had many questions— did sodomy really happened in jail? What was the power pecking order, and did the rapists really get to work in the kitchen?

She would demur. "No, that doesn't happen."

And then when I rifled through her car, I found it: a work manual, with a page devoted to processing sodomy patients.

I went into the house and showed my mom what I'd found, just to see the look on her face.

I laughed, waving the page. "I found it! Sodomy does happen in prison!"

She laughed, too. "My daughter," she said, shaking her head, "my daughter."

We avoided sentimentality. Once my mother and I watched the movie *Glory*. As Cary Elwes and Matthew Broderick sat on their horses, gazing out on the horizon before battle, the music score soared. The moment was poignant.

My mom nudged me with her elbow. "They're going to die."

I stared at her.

"Sentimental people die."

And then Matthew Broderick's and Cary Elwes's characters died.

My mom whispered, "See? I told you so."

My mother would point out a hill topped with snow. And there would be a moment where we would take a sharp inhale at its beauty. This would be followed by my mother's announcement, "Looks like bird poop. So beautiful."

My father was a fountain of pithy and ruthless wisdom. In junior high, when I faced pressures to dumb out and become popular, he pointed at me and said, "You can be popular now, or you can be popular later. I recommend later. Popular later means popular forever."

He didn't think I got the message, so then he invited Lou, the bartender at his restaurant, over to have a chat with me and my brother.

"Lou! Tell them how you were popular in high school!"

Lou was game. He did. He told us all about having girlfriends and going out and having fun.

"Now Lou! Tell them how you're now a bartender! Tell them how that made you a loser!"

Lou was game. He did. He told us, "Kids, don't have regrets. If I had it to do over again, I wouldn't have been popular."

My brother and I nodded. It wasn't until years later that I realized how good a sport Lou had been.

My father taught me Korean words for things like *head, face, eyes,* and *mouth.* Except that he taught me the gutter slang—English equivalents to the word *head* like *gourd, dome,* and *melon,* to *face* like *mug, map,* and *puss,* to *eyes* like *peepers,* and to *mouth* like *cake hole, pie hole, face hole,* and *grill.*

Ten years later, after studying abroad in Korea in a language program, I called my father on the phone. "Daddy! Those aren't the right words for those parts!"

"Yes they are."

"When I said them in class today, the teacher freaked out. She told everyone to forget what I'd said."

My father started chuckling. "And what happened next?"

"She asked me how I had learned the words."

"And what did you say?"

"I told her my father taught me!"

My father was now nonstop laughing. His ten-year-long joke had paid off. "And what did she say to that?"

"She said that I had a very interesting father."

He cackled.

After the stroke, I lost my sense of humor for a while. Everything was literal. Thank goodness, it did return. My father commented on my weight—while I lost twenty pounds the first year of recovery, I eventually gained it all back. And he made note of that.

I responded, "Daddy? Am I perfect to you in every other way?"

"Yes," he said.

"And if I were skinny, too—then I'd be superhuman?"

"Yes," he said.

"Then," I said, "G*d had to be fair."

My dad slapped his knee. "Oh yeah! Fair. Fair!" And laughed. The topic never came up again.

I had come to expect pain, much as someone with an un-diagnosed learning disability might come to expect academic failure. There was pain involved with any physical effort. Migraines if I lifted weights. That sickening aura, the disori-entation, and then the debilitating headache. The dull, persis-tent thud and the knifing, pulsing pain and the nausea and begging for someone to slice my head off my body. Because I wanted to be rid of my body. I didn't want the body I had. I wanted to be just a head.

I read books. I wrote.

I struggled and persisted with my unknown deficit. Alti-tude sickness at five thousand feet instead of everyone else's eight thousand. Joking that I was leaving a trail of crumbs like Hansel and Gretel—that if we were to get lost, we could follow my vomit trail all the way back out. Hiking at the back of the group and learning to vomit with subtlety. Not know-ing this was altitude sickness but accepting that this was my normal. Still stepping forward.

I kept reading books. I kept writing.

I had come to expect pain in my relationships. I once told a boyfriend, "I would rather be loved under painful conditions, be loved in an unhealthy relationship than to not be loved at all."

He said that was fucked up.

I said it was not.

But it was. It was fucked up.

I expected the same devotion, that stoicism in the face of pain, from lovers and friends, and I didn't understand when they did not return undying loyalty to me. And so I became the most disloyal person of all. I hurt everyone before they could hurt me. In my early twenties I cheated on every boyfriend before the inevitable end of the relationship.

When I got to college, I discovered sex. It was the only way in which I could feel good in my body. The only way my mind could turn off. It didn't matter if the sex was rough or gentle or playful. I enjoyed it all.

My body failed me, and there was only so much my heart could take. And so I built a thicker wall. I internalized my anger. Everything was my fault. It was my fault that I couldn't outrun the monsters when they finally outran me. It was my fault that I had headaches; I couldn't handle stress. I hid my vulnerability. I cried into a pillow. I turned off the lights and took an Excedrin PM when I got migraines. And I had sex whenever I could. To run away from my pain.

And there is only one way out of pain. But you have control. You have control over your exit, and for someone like me, that control and exit from pain was attractive. Attractive, like an abusive lover who smells like fruit and cigarettes, and can alternately turn you into butter with his touch or turn you into a ghost when he twists your arm behind your back and pins you until you relent, when he starts out pretending to attack and rape you, and he takes pleasure in doing so, and you take pleasure in disappearing from the room, because you are separated from your hateful body at last, and you take pleasure in seeing your body punished, in seeing it ravaged, and your mind goes very far away, it finally gets to flee.

This is not a good attraction.

But when you are not breathing correctly, when each breath does not give you enough oxygen, you get used to suffocation.

You learn to slowly die.

So it goes.

When I graduated from Berkeley, I didn't walk in my graduation. I watched the English commencement from the back of the Greek Theatre in the rain, not knowing which students in the hundreds of black robes and caps were my friends. I didn't feel like celebrating. I'd finished college, but now what? Who was I if I was no longer a student? It felt more like a funeral with everyone in black robes and the iron-gray sky seeping rain. The cheers and congratulations seemed incongruous to me. It was a recession. Most of my peers did not have jobs, either—and if they did, they were not going into any career related to literature. It was the end. The end. I went home and shut the door, and I did not leave my apartment for a month. At one point, my ex-boyfriend came over to visit. He gave me a giant zucchini his coworker had brought in from a vegetable garden.

I don't know why he gave me the zucchini. Maybe he didn't want to come over empty-handed.

I held that zucchini. It turned warm. It turned soft. It began to wither. Together, we sat on the couch, watching light turn dark turn light turn dark.

A friend came by to visit. He took the squash from my hands and chopped it up into little pieces. He sent them down the garbage disposal.

I wept.

"Please, Christine," he said. "Get some help."

"I don't need help," I said.

"Yes," he said, "you do."

And then I fell in love. I fell in love with a boy who was smart, and who I thought was amazing. He was the smartest person I'd met. I married him.

It was easier to invest in other people. To experience victory through others' achievements, because I didn't have the stamina. I tired too easily. And when I got married, I put my energy into my husband. In particular, I nurtured my husband's career. His achievements were mine. His victories, mine. I filled that hole in my heart with my love and dedication to him and to my marriage. A fortuneteller once told me, "You are two rivers. Together, you make an ocean." I believed her.

12

I love fog. I always have.

I love the way I can hide in the fog, and the way fog obscures the horizon and makes the world the size of a living room. I love its cloudy chill and I love the way it blankets and quiets a city. It is the weather condition under which I write most comfortably.

I understand that it is unusual to love fog. It may provide cover but it also obscures visibility—cars must slow down in fog to avoid accidents, and lighthouses were created to let watercraft know land's end. Lighthouses mark dangerous coastlines and reefs and safe entry to harbors, especially when fog makes it hard to navigate water. Fog is dangerous. In California's Central Valley, tule fog can get so thick, visibility goes down to ten feet—treacherous when driving the interstate at night. It is chilly and uninviting to most. In San Francisco, fog makes summers cold, sometimes colder than its winters.

I moved to Berkeley for college at the age of eighteen, and on the first day in town, my parents helped me move into my dorm. It was a day, like so many days in August, that was enshrouded in mist and cloud; from my seventh-floor room, I could barely make out the hills through the fog, which poured over San Francisco and swept the Berkeley landscape like cotton through a pinecone. Having spent the greater part of my childhood living in the hot and dry San Gabriel Valley, I found the fog a thrill and a relief.

And so were the first six months of stroke recovery. To not see too far. To cozy up. To feel a little lost. To have my senses dulled.

Under IV sedation during my PFO closure procedure, I was not wholly conscious but not unconscious, either. Thanks to Versed, I could not remember a thing, and thanks to fentanyl, I was in a fog, anesthetized against pain.

But the fog was lifting from my life, even though I did not know it at the time.

In so many ways, my PFO closure was a milestone at which I began to wake up. My recovery turned a corner. It was the point at which Adam and I began to diverge in our life goals, even if we didn't realize it then. It was the point at which I began to allow myself to feel pain, to shake off the anesthesia of the previous twenty years of my life.

Anesthesia merely enables other medical procedures to be performed—in and of itself anesthesia does not heal or fix, even though in rare cases it can cause death. Local anesthesia, the mildest and thus least risky form, numbs a small area, like a tooth or finger. Regional anesthesia blocks a large section of pain in the body by interfering with nerve impulses between the body and spinal cord, such as epidurals during birth. General anesthesia renders the patient unconscious, unmoving, and un-remembering for more complicated surgeries by completely suppressing activity in the central nervous system. IV sedation keeps the patient conscious and sedated while blocking out pain and memory during such procedures as colonoscopies and PFO closures; it prevents the transmission of nerve impulses between the thalamus and the cortex.

Some people are afraid of anesthesia, and understandably so. A person is brought close to death, to the deepest level of sleep resembling coma, under general anesthesia. And then brought back out. A woman is numbed from the waist down with an enormous needle taped to her back until her child is born—a friend of mine planned on an epidural but changed her mind once she saw the needle. She gladly bore the pain of childbirth rather than subject herself to a needle in her spine.

Some patients do not mind anesthesia. Some patients embrace it—the idea of pain is too much to bear. More often than not, pain impairs the ability to undergo complex procedures and, furthermore, inhibits healing. We willingly submit and count backward until we march into a deep sleep. Or we breathe a huge sigh of relief once the epidural kicks in.

Anesthesia dampens feeling. Feeling is not always comfortable or tolerable.

Anesthesia is available beyond the surgical suite. Money. Food. Alcohol. Drugs. Denial. Forgetting. Obsessions. Compulsions. Even pain itself.

Adam used money to anesthetize himself in the wake of my stroke; he went on major shopping sprees to mitigate his pain. He was my caregiver. He had lost both his mother and

me. I was still there but unable to console him. I was unable
to maneuver his grief. Our home filled with material goods
made of wood made of fabric made of glass made of steel
made of wool, and the garage with a new car made of metal
made of fiberglass made of glass made of leather. Our home
turned into a museum, filled with delicate glass objets d'art,
fragile and hard, pieces he inherited from his mother, who
collected glass on travels, who bought these pieces because
in her mind she was replenishing the things the Nazis took
from her family during the Holocaust even though they were
not the same things, because she, too, wanted to fill her loss.

Money insulates us from pain. It makes us *feel* a little
less, makes consequences feel a little farther away. And often
money is a form of power, used to fend off helplessness. Even
though money is hardly ever the cure.

In the past, I'd fought off pain with pain. I cut myself.
The physical pain was a way to distract myself from psychic
pain from loneliness from not ever feeling good enough
from pressure to get straight As from failure when I got a
B+ from fear of abandonment from deep rage from lack of
sleep from always having to say the right thing from the in-
flexible demands and expectations of my father. When I got
migraines, I would bang my head against the wall, over and
over again, until I developed a bruise on my temple or on my
forehead. That pain was new, that pain interrupted my migraine; I could not get Adam to understand this even when
he asked over and over again, "Why are you doing this to
yourself?"

I'd fought off emotional pain with food. With the pleasure
of texture and temperature and sweet and sour and salt. I

filled myself until I could fill myself no further, and then I would vomit and eat some more until the pain of engorgement reminded me of my lack of control and failure and added to the ongoing pain.

I'd fought off pain with an attempt to control all things—with obsession and compulsion. With ongoing thoughts and anxieties and then rituals to fend off the anxieties and thoughts: did I wash all the doorknobs did I bring my gloves did I bring my Purell please hug me instead of shaking my hand oh yes I am a hugger but not because I want to be close to you but because I don't want to touch your hands because everything around me is so irreparably stained and I am stained and I cannot get myself clean enough and the world is so full of hurt and confusion and it is touching me and I cannot get myself away from the world so I will clean all the germs off my body where is my Hibiclens soap why is nothing perfect everything is so out of my control everything is teetering. When everything felt out of control I felt as if all my personal space, all the air in the room, and all the thoughts in my head condensed down to the minute details—to microbes on my skin, and whether or not the bread clips were on the bread bags, because that, I thought, that was all I had left, this very small space in the world where all that had to be done was to keep the bread clips on the bread bags, and if the bread clips were lost and you had to fold the bag over itself, everything was lost, everything was lost, everything was lost.

With the loss of memory, so much of my immediate pain was gone. So much anxiety requires memory. Pain requires remembering. Forgetting is a large part of pain management—the drug Versed, also known as midazolam, is often used in anesthesia as a sedative and to inhibit unpleasant memories from the procedure itself.

"Anesthesia without Versed is cruel and unusual punishment," an anesthesiologist once told me.

I agreed with him. I'd had eyelid surgery in Korea, and the surgeon didn't use Versed. I remember every moment of the procedure, even though I was sedated and medicated for pain. It was a terrifying experience. It did not matter that I did not feel pain. It mattered that I remembered.

I could not remember. I did not remember how to ride a motorcycle anymore. I loved riding on the back of motorcycles, my hands barely grasping the pockets of my boyfriends. I had a love affair with motorcycles for ten years before I learned to ride one myself. I happened to begin riding a few months before my stroke.

A few months into recovery, I got back on the bike. I put my helmet on. With Adam's reminders, I checked the fuel and the ignition, made sure the motorcycle was in neutral, turned the engine on, and squeezed the clutch. I rolled forward.

I felt terror. I felt the speed and overwhelming sensation of wind whipping through my jacket and engine noise and the overwhelming requirements to shift and accelerate, and I saw the road disappearing beneath me. I could not coordinate my movements, because they did not come naturally to me. I could not remember how to shift. I wanted to go back home. I signaled to Adam that this was it.

Adam sold the motorcycle when I told him I would never ride again. Ever. In those early months I did not want to put my life at risk. And in those early months I should not have been anywhere near a motorcycle.

When my memory did begin to return, we traveled. I accompanied Adam on numerous business trips, to London, Miami, Boston, China, and Belgium. I developed an immense restlessness alongside my chronic exhaustion—I was literally running away from my pain. I didn't want to stay in one place, because staying in one place meant being still, and being still meant being with myself, and I did not want that at all. I did not know what was happening, and I did not accept that I was changing. In travel, the landscape around me changed more dramatically than I, and the distraction was welcome. Being on a plane to London was better than being in the den by myself—the small room I'd known for years and years, the den that was the same as it had always been, the den that was so much the same that it emphasized my new strangeness.

Being in a different hotel room was comforting. Being surrounded by new smells and different sheets and different light, even alone in a hotel room, was welcome, because these things were just as new as I. New settings erased my nostalgia. And it mattered little what I did; it mattered most that I was in a new setting. I would stare at a hotel room wall for hours before I would venture out to Buckingham Palace, before venturing out in the middle of a dust storm in Beijing before the hotel doorman chased after me with a handkerchief over his mouth and ushered me back indoors. I woke up each morning not knowing where I was, and that disorientation was a welcome respite from my internal disorientation. From being truly lost. So I became lost as a choice.

Before my stroke, I blocked so much in my life. I was the queen of compartmentalization—pushing sadness to one side and turning it into anger and ferocity. I did not allow myself to feel sadness and pain. I always knew exactly where I was, and where I wanted to go, and the road toward that destination. I did not share my stories. I asked other people questions so that they would not ask any of me. I took sanctuary in the fact that no one knew me, even though my deep loneliness wounded me—my loneliness took second place to being as invulnerable as I could be. There were no bad memories. No bad stories. Just a shiny appearance. Unbending. Impenetrable. I kept running and charging forward.

In the past, the sadness would come forth only when I wrote. Writing was my sanctuary. What I could not say, I could write down. And what I could not say aloud was, I am afraid you will leave me and I am afraid to be alone and I am afraid to be afraid and I am sad about the squirrel that died yesterday when a car hit it and I am haunted by the way it bounced up and down on the asphalt in the wake of trauma and I am not sure it was hopping because of momentum or because it was trying to shake off the pain or replace its pain with different pain and I kept driving and I thought perhaps I should have run it over to put it out of its misery but I was too cowardly to do it and take a life and make a complicated decision that wasn't so complicated after all and I turned back around the block to do it and then the squirrel was gone and I had missed my chance to help and maybe it was still alive but where was it I had missed my chance and I felt regret and I hoped it was dead even though I wished it would not have been dying in the first place. And I wrote about a story about a wonderful and perfect boy who drank bile, really drank bile, and languished in its bitterness, and I did it out of jealousy for all the people who were happy and carefree and accomplished.

I poisoned that boy so I could ruin him because I felt so ruined, and then I sent it out into the world.

When the editor of the literary magazine who published that first story met me, he said, "You wrote that?"

"Yes," I said. And smiled. And bounced. "That's me!"

My wall was intact, even though someone had taken a peek behind it. And there I was, unable to be anything but cheerful. I was hiding, you see.

Once, one of my friends asked me if I was a CIA agent.

"What?" I replied.

"I don't know," she said, "I always feel like there's more to everything you tell me. Like you've got a hidden life. That you have secrets."

I laughed, as I always did after being asked about my wall, so I could fortify my wall. "If I were a CIA agent, wouldn't I be in better shape?"

She said, "No. And it's those kinds of replies that make me think you might be a spy."

I laughed again. "I'm not." And because I wanted my friend to not feel like she was insane, I added, "You are not wrong, either." She was smart, because it was true, and she saw what others had not seen.

I never let people in. I didn't even let myself in.

But now, in the wake of the stroke, as memories flooded me and my abilities were not fully rehabilitated, I cried. I raged. In real time. I had lost my emotional fortitude. I could not hold my feelings back. And I did not like what I saw and felt. And to tell the truth, neither did Adam, whose equanimity I admired. Whose equanimity required everyone else around him to be level, too.

I was not level. I was sick.

I was vulnerable. I was healing. I was scarring. I was scarred. I was in pain. I was lost. I was scared. I did not want any of these things. I wanted to be invincible, as I once felt I was.

But I was becoming human.

By traveling, I changed my environment so that I was in a place where I could not be personally challenged, even if I was brought to physical exertion. In an airport, no one would ask me how I was feeling. In a hotel, I was alone and people would only arrive if summoned by telephone. In Hyde Park, I sat alone on a bench and no one knew me. I was no one. And I preferred that to being me, at least at this point in my recovery.

Under IV sedation, also known as twilight sleep, during my PFO closure I was conscious but sedated and un-remembering. The thalamus plays a significant role in pain. It is a chain in the information relay that tells you that what you feel is pain, that pain is danger. That you must recoil. That you must get away. That you must flee. That what is happening is very wrong.

The fentanyl and Versed fed into my bloodstream inhibited the transmission of nerve impulses between my thalamus and cortex and thus prevented anxiety and the creation of long-term memories. In other words, I was in no pain and I remembered nothing. I drifted in and out of sleep, in and out of time, in and around different settings. I talked to my cardiologist under twilight sleep—this I only know because he told me so.

And as much as I wanted all my abilities back, I chased

that sedation in real life as I reached the midpoint of recovery. I did not want to wake up until it was all over.

In the midst of our travels, during which I switched time and place and space and kept myself disoriented so that I could not orient to my new self, we stopped in Tahoe.

I was at the site of my stroke, in the parking lot of the hardware store in South Lake Tahoe, again. It was July, six weeks after my PFO closure. The hardware store no longer had snowblowers out front—instead, green lawn mowers sat in military rows; it had changed, and so had I. A warm breeze tickled my neck, the snow long melted. I felt different. Everything was different. And yet I was there. I felt out of sync, unstuck in time again.

I had no control.

No control whatsoever.

"Here we are," I said to Adam. "I don't feel as much as I thought I would feel."

I had expected an epiphany. I had expected grief. I had expected something. Closure, maybe. But I stood there in the parking lot in my shorts and T-shirt and sunglasses and I felt strange, yet again. Out of place. It made me itch in the dry mountain air. I thought of picking my fingers I thought of cutting myself I thought of grapevines, perfect in lines.

There was the hardware store. There was the parking lot. There was the angle of light. It was no longer winter, and the mountains were brown and scorched, with burned pine trees that looked like black toothpicks. The forest around South

Lake Tahoe had burned down. The landscape was so different, I could no longer relate to it. The place I remembered was no longer the same. So it goes.

The Angora fire was an enormous fire started by an illegal campfire. It took sixteen days to extinguish and burned thirty-one hundred acres. The air smelled like a fireplace. I had smelled this before.

The 1991 Oakland firestorm happened during my freshman year at Berkeley.

A fire started on Saturday, October 19, in a canyon. The fire department put out the flames by that night. But the fire was not all the way out. The next day was hot and windy, filled with offshore winds we in the Bay Area call Diablo winds, similar to the Santa Ana winds in Los Angeles. The smoking land threw out an ember much like my clot, and it caused another fire.

The fire from that ember started at 11:15 A.M. By the afternoon of October 20, the hills were burning in earnest. In the middle of the afternoon, my plane touched down at the Oakland airport. I'd been visiting my parents in Los Angeles for the weekend and was returning to school. There were dark clouds, even at the edge of the bay, miles and miles from the fire in the hills. The fire had taken over a thousand homes by then. So it goes.

My friend and floor mate Lola and I ran into each other on the BART train. Neither of us knew what was going on. But by the time the train emerged from underground at

MacArthur Station, two stops from downtown Berkeley, the sky was black and the sun an orange dot. The train slowed—I would find out later that we were on the last running train. Lola and I had never seen anything like this before. We had never smelled anything like it, either—relentless smoke. When the wind blew, instead of fresh air it blew in thicker smoke.

We were glad to have each other as we navigated the streets of Berkeley up to our dorm.

"What's happening?" we asked someone on the street.

A homeless man said, "There's a fire in the hills!"

We had no idea the extent of the fire until we reached the dorms. There we watched the fire from the seventh-floor balcony, which faced east into the hills. Until the weekend of the fire, it was a largely isolated lounge, because the sunsets occurred on the other side, facing west. If the Claremont Hotel burns down, we were told, we would be evacuating. We were otherwise on lockdown. We watched the tower of the hotel from our rooms, making sure the white-blue lights of the hotel did not suddenly become fiery orange.

My friends who lived in the Oakland Hills arrived. They'd thrown everything into the bathtub and then run for their lives. They didn't know where else to go but the dorms, to my room, knowing this place would be safe.

"Look at my eyebrows!" said one of them. His eyebrows had been singed during the escape.

The fire jumped two freeways, destroyed more than thirty-eight hundred homes, and killed twenty-five people. By the next week, the hills were scarred and smoking. Just like in Tahoe. So it goes.

I now live where the fire stopped, one house away. Across the creek, the homes are new, built after 1991. On our side of the creek the homes were built in the 1920s. When there is a hot wind, we all sniff. When we smell smoke, we get nervous. In this way, we are deeply scarred. Until the PFO closure, I had been afraid of stroking out again.

The fire brought up scars. Like the Tahoe visit brought up memories of the Oakland fire, because memory brings up past trauma, the Oakland fire brought up trauma in my dorm mates. A friend, not a close friend but not an acquaintance either, shut me in a lounge with him during the second day of the fire. It was the first-floor lounge, and it was deserted, because the first floor had no view. He said that he'd just tried to kill himself. He said that when he was a child, his father had locked him inside their home and set the house on fire. He said the world was on fire again. I was eighteen years old and said I understood, but I did not, then took the first chance I had to unlock the door and flee the lounge. That friend killed himself four years later by throwing himself off the math building.

There are so many chances for healing that I did not take. There are so many chances for healing from which I fled. There are so many scars from a fire. From an ember. From a memory.

Twenty years later you cannot tell there was a fire in the hills, other than the swaths of newer construction. The authorities put in standard-sized fire hydrants. There would be a fire again in 2008, but it would be put out. Immediately. And there would be no remaining embers to cause another fire again.

And in Tahoe the forest would rebuild but be different. Gymnosperms released their seeds. There would be regrowth. In fact, that is how pine trees are programmed to survive. Pine trees require destruction for new growth.

There is plasticity in nature.

I was not the same body. I was not the same person. Parts of my brain had been erased. Parts of my brain were new. The hole in my heart had been closed.

I was under anesthesia. Twilight sleep. Jumping back and forth in time. Weaving lessons together. Weaving narrative and story and deep memories. Weaving self. All while numb.

When I woke up, so much had changed.

13

It is 2003. I'm thirty years old. I do not know I have a hole in my heart. I have not yet had a stroke. I am gasping, as I do on backpacking hikes. There are miles to go, but I think of the trail ahead not in miles but in terms of what I can do in ten-second intervals. I put one foot forward in turns, sometimes just an inch at a time. Adam, Mr. Paddington, and I are walking along a piece of coastline too steep and rugged to build a road on, resulting in the only part of California's Highway 1 that veers inland. This trail is called the Lost Coast, because of this very fact. And then we stop. The trail is blocked. We do not say a word. In front of us sits a Roosevelt elk. A Roosevelt bull elk. It will not move. And it is enormous—when it stands up, its shoulders are taller than I am. Around us is brush and greenery, much of which includes poison oak. We stay there for a while, the elk and us, until the elk gets up, its antlers like a fascinator headpiece. We back up, and since we have no choice, we pick our way off the path and hike a wide berth around the bull.

We are following a map, but the map never fully prepares you for these encounters, these surprises. There is no spot on a map that will tell you, "Here is a nail on the road. Do not drive over it. Here is a large elk. Hike around him. There is an agitated wasp on this bench. Be careful. On this day you will have a stroke."

Mr. Paddington and I have been friends for three years at the time of our Lost Coast hike. This is not our first backpacking trip together, and it is not our last. We will become the best of friends over the next few years. In ten years he will save my life and I will save his, and a year after that, we will fall in love. We do not know this while on the Lost Coast in 2003. We hike north. We camp on a beach. We stay awake all night in our two separate tents, knowing that we have pitched our shelters above the tide line but still anxious about the sound of encroaching water. In the morning we wake up to a blushing sky and a black sand beach before we hike back out. On that trail, on the last two miles back to the trailhead after two nights in the wilderness, we are in pain. Mr. Paddington and I don't know if we can go on. We do not say so aloud, but we share a tongue-biting silence and the clipped walking stride of people in distress. Our feet are burning, because we have been breaking our speed down the steep descents with our feet for days. The water in our Camelbaks is low. There is no more shade ahead, and we know this, because we walked in on the same trail two days prior. Sometimes it is worse to know what lies ahead. To our right is the Pacific Ocean and ahead of us, miles of yellowed grass. When we hiked in, we took pictures, the ocean to our left through gymnosperms permanently bent from consistent winds, then meadows on the sunny southern side of gorges and prehistoric fern grottos on the northern sides. When we hiked in, the grass was uphill and the ferns downhill. Now, capturing this landscape on our cameras is the last thing on our minds. Adam, meanwhile, runs ahead, because he can. His ectomorph body has longer legs. We, the endomorphs, plod on.

The Lost Coast Trail is a one-way trail that most people hike in one direction from north to south; we chose to walk in from the south and turn back around at the halfway point. Adam, Mr. Paddington, and I walked a total of sixteen miles. We climbed steep ridges, two 1,100-feet ridges and one 800-foot hill, twice—we gained and lost a combined 5,503 feet of elevation within 8 miles. The slopes were often eroded and slippery. It was beautiful. Until pain overcame beauty. And then we reached the final plain, at which point I considered crawling. There was nothing to do but move forward, even if slowly, even with rest breaks. Adam with his lengthy stride has long left us, and he is likely already in the car, waiting for our return. I think of air conditioning. I think of upholstered car seats. I think of ice water I think of ice cream I think of a bath I think of clean clothes I think of a hamburger.

Mr. Paddington sighs. He turns around and starts walking backward.

I stand, shocked. By his creativity. By his audaciousness. By the fact that Mr. Paddington has chosen a new way to walk.

It is harder to stay still than it is to move; my feet shriek where they stand. I shift my weight from one foot to the other.

The adrenaline starts to kick in. I howl. I start laughing uncontrollably.

I scream, "You're a genius!" The elk scatter.

And I too start walking backward. There is new pain in new movement. And this new way to move is a relief.

Now the endorphins kick in. For two more hours, as we approach the car, as we drive out onto the highway, as we order fast food, I laugh with abandon.

Listen—I am recovering from my PFO closure in the summer of 2007. It is month six of recovery. The first six months of healing were dramatic and the pace of recovery exponential. The last eighteen months, less so. The last year and a half is about waiting and wading through the grass, taking each step to the end. The last 20 percent of my recovery takes 80 percent of the time.

I'm thirty-three and I've just been released from the hospital. I'm learning to walk backward. I am tender but repaired.

By the time of my PFO closure, I felt 80 percent recovered. I could do most things, so long as I made sure to rest afterward. I could read a short story. I could write longer and longer blog posts. I could assemble a peanut butter and jelly sandwich. I could coordinate an outfit without feeling like my brain was melting. I could even choose something other than a hamburger off a restaurant menu. In fact, I could do all of the above inside of a day. It would take me longer to do so than before the stroke, and each task came at a cost, but I could do them.

I would still have setbacks, though. If I pushed myself too hard—say, went to a department store and then went to lunch with a friend in a noisy restaurant, I would be near collapse with disorientation and fatigue. In fact, I would have to sleep and rest the next day, too. Those were times that felt like two steps forward, three steps back.

It still is that way. If I am at a high-decibel restaurant and I am already tired and I'm with people I don't know, I want to clap my hands over my ears and close my eyes and stop talking. This is not the way I used to be.

But to the outside world I looked like I was in pretty good shape. In fact, I was in pretty good shape, given only six months had gone by. I was functional. Functional enough to technically survive the rest of my life as I was. But I was also well enough to know that I was not fully recuperated. I wanted to thrive. Surviving was not good enough.

In the wake of my PFO closure, I was brought back to the beginning. Marked exhaustion. The smell of hospital on my body and my clothing. Lingering bruises from the Lovenox shots I'd taken in the days following my TIA. Sticky adhesive marks from circular EKG and telemetry stickers, graying with body grime. At the entry point of the catheter, a tender mark. For some reason, I couldn't bend—bending down took my breath away, so I had to ask people to pick up a fork I'd dropped. I wasn't allowed to carry anything weighing more than ten pounds. So I had to ask people to carry my bags. In this way, I was forced to ask for help. And when I laughed, I had to hold my groin—I felt like my incision would burst.

But this time I regained ground quickly.

My PFO closure was the point at which I no longer considered myself sick. I would not have another stroke, at least with the same root cause, ever again. It provided me with immense peace of mind, like reassuring an earthquake-prone region of the world that there will be no more earthquakes, ever. I began to trust the earth on which I walked.

———

Vilfredo Pareto discovered the 80/20 principle; in 1897 he observed that 80 percent of the land in England was owned by 20 percent of the population. This rule has since been applied to many aspects of modern life. For example, in sales teams, 80 percent of the revenue usually comes from 20 percent of the group. In time management, the first 80 percent of results, it's said, takes 20 percent of the effort. The last 20 percent, it is said, takes 80 percent of the overall effort.

I had reached 80 percent in my recovery.

What was most challenging at this point was bridging a divide between what my brain was doing, what my body was doing, and what my mind was doing.

My brain was healing itself and taking on new ways to do the old things. It was bringing back function and building new neural pathways.

My body was rebuilding—getting fit, reveling in fully oxygenated blood.

Before my stroke, I defined my limits not in energy but in time. "I don't have enough time to do all this!" I used to say. But now it was "I don't have enough energy! My mind is melting." Whereas before my mind could take charge over any matter, could push my body past its limits, pushed my body to backpack and hike, my mind could no longer keep up. My mind was reeling—my mind could not wrap itself around this healing body, this new identity, this new facet, this new thing that suddenly worked. My mind, on which I had so strongly and solely relied for so many years, was at a loss. It compensated. It told me I couldn't come back. It missed my old life. It defined a comeback as going back in time. It reached and stumbled and despaired. It did not understand, and so it made up stories to fill the gap, to make sense of black scars and deficits and fears and hopes. You will never be the same, Christine. You are now a body without a mind. You can now breathe, but you cannot remember the things you need to remember. You cannot balance a checkbook. You cannot read a novel. You cannot write a novel. You cannot keep back your tears. You cannot keep back your anger. You cannot read a map. You are a zombie. You cannot engage in more than one activity a day before collapse—on days you want to write anything down, you must not eat, go outside, or shower before you get your words down in your journal, because otherwise your mind will not have any words at all. You cannot bake a cake, still. All this happened.

But my mind and body, long at war with each other, would come to a great peace beyond my imagination. The thalamus, the part of the brain responsible for relaying messages between the brain and body, had broken. My mind realized it needed my body, needed to recognize its new and strange strength.

My doctors cleared me to exercise six weeks after my PFO closure. I hadn't exercised in more than six months. I expected to feel my lungs burn and my heart to thump out of my chest when I got on the treadmill and cranked the speed up to four miles per hour, my previous maximum. I readied myself for the inevitable agony, the thing I associated with exercise. I readied myself to feel worse than I had before the stroke. Because everything had been worse to this point.

It was easy.

It felt like walking.

I double-checked the machine. Looked at my feet to see if they truly were moving at a faster clip. I could not believe the absence of pain.

And so I cranked it up to five miles an hour. I jogged.

It was easy.

And it was not just easy—running felt better than it had ever felt before. Every step was no longer a struggle. I understood how running felt like freedom. I was not gasping, I was taking deep and measured breaths. It was, I kid you not, as if with every breath I lifted my body off the treadmill. I no longer felt the immediate pain I'd always felt while exercising.

I cranked it up over five miles an hour and did a slow run. Left and right, up and down, breathe in and breathe out. Count to ten, look up. Count to ten, look up.

I ran for half an hour this way. I was astonished. I had never done such a thing before, let alone after having sat on a couch for eight months straight.

I got off the treadmill, sweaty and invigorated. Actually sweaty, because I could exert myself. Actually invigorated, because it felt good. I sat down on the floor to stretch, and instead of stretching, I took a breath, and when I exhaled, I exhaled sadness and disappointment and rage and my chubby childhood years and frustration, and I emptied myself until the voices in my head—a lifetime of voices that said I was not good enough that I was too fat that said I must starve that I was not good at sports that I would never be able to run or jump like anyone else for some unknown reason—went quiet. And then, in that emptiness, I cried. I was not sad. I was tired. I was inspired. I was so so so relieved.

I was also so so so angry for the years wasted on berating myself. It had not been my fault, after all, that moving my body was so difficult.

I hadn't needed to struggle with weight all my life. I hadn't needed to struggle with fitness. I hadn't needed to starve myself to be a size 8. I need not have gasped for air during all those previous attempts at exercise.

I waited for the inevitable migraine to develop, something I'd come to expect after any vigorous workout.

It did not.

My body had done something amazing. My body was new. My body had not failed me, for the first time.

And because I could breathe, I could exercise to exertion. And because I could exercise, I got fit. Truly fit. I was changing, now physically.

And my mind—my mind was in awe. Even if it was angry and sad and in mourning and all those other familiar feelings, I was feeling a respect for my body, for the first time in my life. And for the first time ever, my body made my mind feel better. My thalamus communicated and coordinated these efforts.

And then the true healing began. The last 20 percent, which took 80 percent of recovery time, was slow but filled with insight and realization. The last 20 percent of recovery was about becoming a whole person—not just my mind and my heart but bringing my body along for the journey. I had fought so hard the first six months of my recovery with just my mind, because I had no trust and faith in my body.

Now it was my body's turn to lead.

I began to listen to my body, instead of ignoring its needs.

This was a new dynamic, one in which I faltered—my mind was a tyrant, but the thing is, my mind and brain were tired. And my body, now that it was oxygenated and therefore could be fueled, was the strongest member of this team. The part of me that had never been reliable was now the most reliable.

I stopped punishing my body. I exercised. And because I needed strength, I made sure my body was strong. My body did not let me down. When my mind would tire, my body would hold me up. It would not let me down again.

I began valuing my body for its resilience and strength and stopped devaluing it for its appearance. It had too long been my foe—it had been overweight, infertile, exhausted, depleted, and it had too long been a hurdle to overcome. I had abused my body—starved it, cut it, and neglected it, in an attempt to erase it and put it in submission. But I had been unfair to my body, which had had a defect all my life. Which had been working with a handicap, yet worked so hard nonetheless. Which now could do all things, because it had learned to bear pain for so long.

Because I had been unfair to my body, I had been unfair to myself.

I began to value my own resilience and strength.

My mind began to see the ways in which I had improved over the last few months. That I could even write at all was an improvement. That I could even do one thing a day, alone. That I could travel. That sleep helped.

Yet, as revelatory as my insights were, these developments came slowly. They did not happen overnight. They happened over the span of a year and then for years afterward. As rapid as the first few months of recovery felt, the last year of recovery was agonizing in its glacial pace.

This part of recovery is so private; to everyone else I looked fine, and the improvements I made were measurable only to myself. This is the part where the doctors left me, having said they could not help further. The therapists had said their work was done months before that. My friends had tired of hearing about my illness. This is the part where I still knew there was a ways to go and I was not sure I would get there but I had come this far so I knew I had to keep going. This is the part where I knew I had to reconcile a photographic memory before the stroke and just being able to remember what happened during the course of a sitcom but not remembering the license plate you saw earlier that day was not okay. This is the part where I could not calculate the tip on my meals and other people told me they could not do this either and it's okay, but it was not okay because I used to be able to do this. I used to do more.

This is the part where I wonder if I would have to be okay with "just okay" for the rest of my life, with this new me, this new, mediocre me. This is the part beyond bones and skin and muscles. This is the part about wondering who I had become, and who I wanted to be, and whether my spark would return. This is the part when I wondered if I would continue to be a golem or if I would become a human again, because really, I felt so unlike myself, inhuman and separate and strange.

It's not that I wanted this part to be private. But it had to be. I wanted to share with everyone what I was going through, but I could not articulate it at the time. I went quiet. I wept. I rallied. "I'm almost there," I said. And I would leave it at that.

I had almost zero auditory learning capability and had to write everything down. I was able to put up a good front—because I learned to rely on intuition rather than on cognition; I felt like everything was a guess that turned out to be correct. Like seeing a face and not knowing the person's name, but turning to the person and thinking she *looks* like a Polly because something in my gut is saying Polly Polly Polly so I'm going to say Polly is your name Polly? Yes. But this became my new relationship to memory. Still, to this day, I remember people's names in such a visceral way, one that feels uncertain but, over time, has become dependable. It feels like reading minds.

Along with being able to remember, and thus to grieve, came nostalgia for the past—like listening to Tori Amos and remembering afternoons in the college dormitory or a concert I once attended with a boy who once loved me. It was a relief to feel nostalgia in this beautiful way. I still had problems with words at times—I'd search for words that would describe the metallic smell when rain falls in the summertime, and it would come out, "when rain hits summer." I'd still switch homophones, my most common and lasting aphasia—earlier in this chapter, I thought I wrote "Our feet are burning," but I'd really typed out "Our feet our burning."

The interstitial spaces are hard. But I came to occupy them. And then they became my new space.

I look at the way I treated pain then versus how I treated pain in the wake of my stroke. I would go on to yoga and learn to breathe in a way I never could before. I would learn to channel the ability to breathe into a new breath. Tara Stiles at Strala taught me this—to breathe and fold over to breathe and stretch out and hold and breathe and look ahead and breathe and inhale and exhale and feel my feet on the ground and my arms in the air and feel my strength and breath and then one day to lift myself up off the floor with my breath.

And then I would have a baby. And I would breathe through the birth.

And I would encounter heartbreak. My marriage would end. And I would breathe through that, too.

And through all of this, I would let my body lead, because my heart and mind were breaking down—from birth, from postpartum depression, and from heartbreak. And I would let my breath lead, because this was a lesson I'd learned in the wake of my stroke. I would breathe so I could feed my body. I would feel the sadness in my bones. And my mind, because it trusted my body, acquiesced, allowed the pain to rush through. I have to let it run its course,I thought. I knew that pain and sadnesss and hopelessness were all temporary; healing involves inflammation and pain and struggle and purging

And like my time on the Lost Coast, I knew the last mile is the hardest. It is the mile you often travel on your own. Where you must go outside your comfort zone. Where coping is finding a new way to do the old things. Your doctors are no longer there, your friends think you are fine, and you are functional enough to not elicit any sympathy. It is the last minutes of childbirth. It is the end of a manuscript. It is all up to you.

14

By the end of the first year, my stroke was very old news. I could run and exercise, so everyone thought I was healed—thus I had little to complain about. Despite stroke being a physical affliction, my deficits were neurological, mostly noticeable only to myself. In those last months, things were even more invisible to the outside world.

I did have fewer complaints. But I knew I wasn't all the way back. I knew there was still a ways to go until I could lead the life I wanted to lead. I still could not write fiction. I still forgot things on a regular basis. And for all my improvement and the remaining setbacks, I was easily exhausted. But who wanted to hear this? Who wanted to hear about my discouragement? And I didn't want to hear that I should be grateful. I wasn't. I wanted to be better. Truly better. Better than before. Not better as in better than the month after my stroke. So by the end of the first year, I fell into a deep silence about my illness and on-going recovery, replacing it with idle chitchat. My horrible wall was back up.

Behind the wall, I stewed and I waited and I fought.

I was always tired, even upon awakening. Every element of the day, even after a year, involved a trade-off in my over-all energy level—brush my teeth and wash my face and then write in my diary later, or brush my teeth and wash my face and do my hair and prepare breakfast?

Regardless of what I did in the mornings, I was knock-out tired by early afternoon.

In the afternoons I reeled with exhaustion. I could not think straight. I could not see the items on a lunch menu. At work, the office sounded like a carnival, even though all that was happening was discussions about big data and load balancing and marketing collateral and maybe the odd an-ecdote about Halloween costumes by the water cooler. There really was a water cooler. When the water belched out each cup, it sounded like an ocean swallowing an island. There was no shouting—in fact, everyone was actually talking in hushed voices because it was an office after all—but it was all noise to me. Everyone ate lunch around the conference table, and the waxed-paper sandwich wrappers made loud staticky noises and I clapped my hands over my ears and closed my eyes and when people asked me why I just said, "Oh, I have a headache," because there was no way in that office that I wanted to tell people that it was too loud and there were too many colors and my brain was overwhelmed and I could not take it anymore and all I wanted to do was shut my eyes and go to sleep. There was a lactation room at the office, and sometimes I would go there and close the door and sit in the dark with the lights turned off. One day, my boss found me there.

"But you look just fine," I would be told.

Yes, I did. I looked just fine.

But I'm not, I wanted to say. I am so far from being fine. Instead, I shrugged. "I just needed a break is all."

I knew telling the truth would have sounded like a broken record. And I had no idea what "fully recovered" would look like—I had to try to be happy with wherever I was, because any moment I experienced might very well be the end of the journey. I might never have total recall again, and I would have to be okay with that. I might never be a social butterfly again, and I would have to be okay with that. I might not be able to finish my novel, and I would have to be okay with that. I would have to.

And because I have never discussed this last phase of recovery, writing it down is challenging—what is the story here? It could be a Hollywood ending. It could be a fable with lessons. It could be a Pollyanna-Happy-Wonderful-No-Regrets story. It could be about coming full circle. It could be a depressing story about loss.

That last year revolved a lot around helplessness. I had the same deficits as before, though to a lesser degree. I could hide these deficits fairly well. But because they were the same deficits, discussing them felt like complaining or whining. Because it was. Because complaining is a sign of helplessness.

In the fall of 2007 I reenrolled in my MFA program.

I had one semester remaining. I knew I could at least get myself to class and sit and chat politely. I could read story drafts and offer some sort of feedback. I went back to finish my degree at this time because I thought that this was as far as I would get in my recovery, and I expected very little further improvement. I might as well, I thought, finish up now.

My thesis advisor, Yiyun Li, with whom I had forged a good friendship, knew I could not remember anything. She knew I could not write my novel even though I said I thought maybe I could. I suspect she noted the cadence in my speech, the way in which my diction had changed, and maybe just my lack of confidence. So we changed my thesis manuscript. Together we decided to submit the short stories I'd written throughout grad school, instead of my unfinished novel.

"Let's get you graduated," she said. "The novel can wait."

I nodded. It was a painful concession, but the truth was that my brain could not write a novel. I was grateful that she was matter-of-fact. I made my novel wait.

I did not expect to retain anything that semester. I was just there to finish.

And I do not remember, not to this day. I could not remember where I parked my car each evening, and so I tried my best to park as close to Mills Hall as possible. I remember my brother urging me to get a Handicapped placard to mitigate this problem. I refused.

"I'm not really that handicapped," I told him.

"You can't find your car. That's handicapped," he said.

"I'm not limping," I said. "I'm not in a wheelchair. Everyone will question why I have a handicap permit."

Even as I knew I was disabled, I was in denial.

"Suit yourself," he said.

I do not remember any content from this semester, other than what class I took. And that is because I see the class on my transcript. I took a fiction workshop with Cristina García. The plan was to workshop stories I'd written before my stroke. I do not remember which stories I workshopped.

In fiction workshop, we held discussions that I do not remember, though classmates do.

I remember encountering one of my classmates a couple of years after that semester.

"Hey, Christine!"

I turned around. It was someone I knew. I searched my insides for the echo of a name. What was his name? I let go of all my errands and obligations for the day, and the yammerings of my exhaustion, so that my mind could go quiet. So I could hear the whisper of his name. Scott? Scott. Scott! Scott? Maybe.

"Hi. Scott?"

"Yah! Hey! It's been a while."

It had been a while. It had been at least a year since I'd seen him, since I'd graduated from my MFA program.

"I just wanted to say hi. I'll never forget that workshop discussion on writing sex scenes. We said some wicked things!"

I did not remember. So I said, as I usually did in such situations, "I don't remember. But I totally believe you when you say it happened!"

These are the encounters I have all the time, even to this

day, from people I met during my recovery. I don't remember what happened. It's like a black hole. I have to trust their narratives. It's like I wasn't there, even though I was.

Sometimes people tell me stories that jog my memory. "Remember the time you picked me up from BART when I flew into town from Toronto? And we had *huaraches* and *alambre*? That was the first time we met in real life!"

"Oh yeah," I'll say. "And we also had *huitlacoche,* and that was your first time eating that?" But I will not remember more than that.

And then my friend will remind me further, "You were kind of spaced out, but still you."

"Yes." And I will remember, "I totally forgot where we parked the car, didn't I?"

"Yes," he will say, "but we found it. Next to the watermelon vendor."

But more often than not, I don't remember anything at all. Sometimes friends told me their own stories during my recovery. I am thankful to them. Deeply thankful. These friends knew I was a writer and told me their stories with the intention of awakening the storytelling part of my brain. Not the "Oh, I had a medical situation too" stories or "My father was once sick" stories or "Someone I know also had a stroke" stories, but real stories about their lives, about their families, about moments that meant something.

Like the time my mother flew up to visit me when I finally allowed her to. The first thing she said was, "You are going to write stories again." And then she told me something I'd never heard before.

She told me a new story about my uncle. How my father

had taken an extra job to support the family and how he had sent extra money to his younger brother in Korea. How my uncle had fallen in love with a woman at the time. How he took the money my father had sent him and spent it on a helicopter to impress the woman on a date. In poverty-stricken 1960s Korea, before the 1988 Olympics and high-rises and subways and K-pop. When a dinner would probably have sufficed.

It was romance. It was tragedy. It was brothers. It was yearning. And it struck my heart. And I cried. "How hungry they all were," I said, "so hungry for love, different kinds of love."

"Your uncle is a bastard," said my mom.

Like the time my coworker Garrett told me about a winter vacation he spent working on his family's chicken farm in Japan as a teenager. How the birds would flutter and scratch his arms and how they stank and how the room smelled like fear and shit and how the feathers flew and the wings flapped and how the workers held the chickens upside down, sometimes four, five, six at a time and how the birds all went into boxes, their faces lit up in shock, upside down sideways right side up.

How the rice hulls dried his skin on harvest day.

These images helped me begin writing fiction. I eventually wrote a story about a chicken farm a year later. It did not get published, but I am glad for having written it.

These stories struck my emotional center. In his essay "The Abyss," Oliver Sacks discusses Clive Wearing, a man who can no longer make new memories but can remember his wife. "There are clearly many sorts of memory," Sacks writes, "and emotional memory is one of the deepest and least understood." The stories people told me out of love, about love, went into my brain through my heart—another way in which we are able to remember. This is why, even though we have so few memories from early childhood, we do retain a few of them—the ones that strike at us in the deepest places, the places of fear and love and exhilaration. These are the memories that never leave us. Our first trip to Disneyland. The first time we were hurt—that time in the emergency room when I screamed in a dark room on dark chairs and then the room with bright lights where the doctors pinned me down—when I'd fallen down as a toddler and broken all my front teeth, and how I never wanted to fall ever again. Our disappointment. That time in the snow, when I tried to throw snowballs, but could never hit my target because, unbeknownst to me, I was too small. Those summer days under the El in Queens eating fried chicken and pizza. The one-way plane ride from New York City to Los Angeles, how the strawberries on my lunch tray matched the red seats, and the first time I realized I hated airplane food. Or our first time witnessing violence and fury. The time my father hit my uncle, and the morning afterward, when my mother and aunt tended to their split lips and black eyes.

And now I understand that it is through emotions that things are most permanent in my brain. It is when something pierces my heart that it enters a part of the brain otherwise inaccessible. Happy experiences. Joyful ones. Angry. Sad. Scared. It all matters, I have learned.

There is so much I do not remember. I wish I could. And that is so much a part of my obsession with photography and journaling—I wish never to forget, even if it is inevitable. I wish to remember all the details.

I went back to finish my MFA program because I thought I was finished.

But I was not. I kept writing.

I told my story over and over again—the same vignettes and anecdotes—each time I revised as I gained clarity and understanding. Boring happenings that took on more dimensions with lessons learned. And more and more detail over time, as I healed.

One time I hung out with a Famous Writer for a couple of days. We planned a party together. He told everyone this was happening: "I asked Christine what she wanted to drink, and I was at Target and I bought it."

The next time he told the story he said, "I just knew Christine wanted me to buy a ton of liquor."

Then after that he said, "Oh, man! Christine told me to buy all the liquor in that store!" By now, the party was well under way.

I looked at him each time, bemused. What was he doing? This was the truth but not as it happened. It had happened in such a banal way. He really had called me. He really was at Target. He really did ask me what I preferred to drink. I did tell him what I preferred. He did buy my preferred liquor.

"Aha," I said aloud. "You're the consummate writer. You're revising!"

He looked up. He smiled. I felt like he and I were the only two people in the room.

"Your secrets," I told him, "are safe with me." And I laughed. Because no one else in that room was aware that I'd had a stroke years prior. Because by then I was well. My stroke was in the past. And yet not.

It's a process. And I learned that acceptance and understanding don't happen at the same time—maybe they happen in parallel, but understanding comes a long time after acceptance.

My stroke taught me lessons, many of which revolved around writing.

I learned that memory comes in modules. That narrative comprises pieces that must be woven together to create a unique texture and pattern. They are connected by pieces of thread. I would begin to tell a story because I had an image, and I would describe the image and then I would trail off, because . . . because I'd forgotten what came next, and I would have to remember the next thing and connect it to the first and then I would forget and sometimes I would never remember, and it is all a bunch of pieces.

My brain was blown apart, and it took that catastrophe to understand how it worked. Sometimes you have to lose a part of yourself to know what it is it does.

I had to lose my ability to write to understand writing, fully, as it pertained to me.

Storytelling is emotional. It is about asking people to remember something that meant something to you, and the only way someone can remember your story is if it strikes the heart, at the reader's emotional center. And it must be authentic.

I graduated from my MFA program. I limped to the finish.

And I didn't walk in my graduation. I didn't walk, because just as I felt when I earned my undergraduate degree, I didn't want to celebrate. What kind of life was before me? I'd graduated from my MFA program, and yet technically, I wrote worse than I had before.

I didn't walk because I could not be in a crowd just yet. I knew my brain would melt in the siege of faces and how-are-yous and congratulations and the hands and the robes and the sunlight and the pictures and all the hugs.

I didn't walk because I didn't want to answer questions. I didn't want to talk about my stroke. I didn't want to field reactions.

But I did have my own private ceremony, years later when I visited my thesis in the college library. I wanted to see it bound—because the idea that I'd written a thesis and graduated from my MFA program did not feel real. I did not remember printing my thesis out. I didn't remember buying the special thesis paper. I didn't remember putting it under my professor's door. I didn't remember how it came to be in its physical form. But there it was, in blue binding, with my surname.

I turned the pages. It was as if someone else had written it.

And it struck my emotional center—not because of the writing but because of the journey involved. I'll never forget my thesis advisor at her pragmatic best, helping me get this done. This was the culmination of my work. This was the beginning. Just graduate, she said. Get there.

I told myself I would get there. I would get there in a different way. I would get there a different person. But I would get there.

My thesis felt like someone else had written it, because now I was a different person; someone else had.

I wrote every day throughout my recovery and after my MFA graduation. In the beginning it was nonsense. And then it began to make sense, even if I wasn't writing the same way or at the same caliber as before. My writing had been ripped down to the studs, and it was being rebuilt. And then it began to make a lot of sense. My writing improved, because I was honest, because I didn't have walls, because I had nothing more to lose. Because why else would I be a writer?

My writer friends urged me to apply to VONA, Voices of Our Nation, a workshop for writers of color by writers of color, again. I'd been to VONA before and studied with Junot Díaz, but then the stroke happened, and I didn't bother.

"But Chris Abani is teaching!" wrote my friend Pia.

"So?" I replied. "I can't really write."

"You'll learn so much. Just apply. See what you can do."

And so I did. I wrote not from a technical place but from my heart. I told VONA that I wanted to return to writing, and theirs was the safest place for return. That I wished to put my toe back into the pool with VONA as my setting.

The next summer, in 2008, I attended VONA again, and took a workshop with Chris Abani. By then I could read. By then my fiction writing was slowly making its way back. I was picking my way through the landscape of my novel— trying to imagine new chapters again. I had written a short story—the one about working on a chicken farm in Japan, the one inspired by my friend Garrett.

On the first day of workshop, in the very first minutes, Chris asked us to write a deeply personal anecdote about what we wrestle with.

I raised my hand. Prompted to speak, I said, "I'm a really private person."

I hardly knew the other people in workshop. I didn't know him. I had gone through a war with my body, and I felt tender.

He listened with steady and unblinking eyes, eyes that at one point in his life faced psychological and physical torture in a Nigerian prison and behind which lay a brain that withstood and survived that torture and then thrived. He asked a question that was more than the question, in the softest of voices: "Then why are you a writer?"

I closed my mouth.

He said that a writer cannot be private like that. We must share our truths. We must be brave.

And so I agreed. I picked my pen up and I wrote about the year of my grief; about the loss of my memory after my stroke and how it paralleled the grief from losing a family member the same year. And how my memory of grief in 2007 returned in crisp relief against Adam's fading mourning, and how my grief made me feel alone because that is the nature of grieving but also because I felt out of sync with everyone and the world around me. I had never opened up so much before. I was becoming a better writer. I was making my way toward my goal: to heal my writer's brain.

Recovery is not straightforward. Things move backward and forward. Depression comes and it goes. One day I can read part of a book and the next day I can only pick up *People* magazine. Wasn't I better?

I faltered, I leapt, I ran, I lay down, I got up, I walked, I limped. In the thick of recovery, I couldn't even tell if I was getting better. I yearned for those early weeks where I was so injured I didn't care because I didn't know. But I was getting better, very slowly, so much like hiking up a mountain inch by inch, where you can't see the landscape change very much, when movements are so slight, you might as well be standing still, when the destination is so far ahead, you wonder how it is you can get there at the pace you're going.

Recovery is harder to write, because of this uneven arc. How to let the reader know how it feels to navigate terrain while providing a forward-moving story? How to be frustrated, and not make the reader frustrated—and to glean lessons learned and portray a kind of resilience without glorifying yourself? It is different from the sickness, which is something that happens, a flood of descriptions and confusion and a definite downward slide—initial affliction is a more exciting and cohesive narrative.

Recovery is a day in, day out onslaught of alternating perseverance and self-pitying tears. Was today better than yesterday? Sometimes it was not. Was this week better than last week? Sometimes it was not. Was this month better than last month? Probably. Wake up each morning and ask, "Is this good enough?" Because you know that this might very well be the full extent of recovery.

Recovery is repetitive. Keep doing the same things over and over again. Fail all the time, react to the failure, learn from it, feel from it, move on. Do it again. Fail some of the

time, react to the failure, learn from the failure, feel the failure, and then recognize the successes, react, learn, feel, move
on. Each time, different reactions, different lessons learned,
different emotions, and then moving on.

Go to occupational therapy. Write words.

Go to therapy again. Write a small sentence.

Go to therapy again. Write a longer sentence.

Wake up.

Wake up.

Repeat. Repeat. Repeat.

———

In the landscape of chronic illness—the kind that lasts over
six months, over a year—there isn't much medicine. If you
suffer from chronic pain, no one wants to hear it. If you are
sick for longer than a certain amount of time, people grow
weary of that narrative. Sometimes you take on the mantle of
being a Sick Person, maybe willingly. You forget how to elicit
sympathy and empathy. So you demand it, like I did.

Your illness is in your head. You are imagining it. You are
making all of this up. You are not that sick. You are just fine. Just
fine. Please be fine. I don't want to hear anything else from you.

And then one day, suddenly, you are fine.

And you remember the book you read on the day of your stroke. The one that has entered the deepest part of memory. The book that becomes a part of your emotional landscape. The book that you will always equate with your stroke and recovery. *Slaughterhouse-Five.*

———

I shook off the cloak of Sick Person. After being ill for more than a year, how does one leave that identity behind? I couldn't help but miss it, just as when I lost weight, I still identified with being fat, because I'd been fat all my childhood and for most of my adult life. And yet I so wanted to be rid of illness. I was still staggering.

"I get exhausted in the afternoons," I wrote, "and dreadfully sleepy, only to revive (nap or not) by nighttime. It feels like jet lag. Have I been traveling to other dimensions in my sleep? If so, it must be to wherever three o'clock here is bedtime elsewhere. Europe, perhaps?"

In this and other ways, I was no longer the same—but did that mean I was still sick? I had friends recovering from cancer who were declared to be in remission—and yet, one isn't "cured" from cancer until five straight years of remission. The specter of illness looms. When I reread my diary, I am reminded of my friends who have also survived illness, who found it hard to shed the identity of Sick Person.

It was a struggle to let go of the old me and embrace the

new one. Part of healing was inhabiting change and my new body and brain. For as long as I mourned and struggled to turn back the clock, I wasn't going to be well.

When was it that I let go of the old me?

I retold my story over and over to myself and to others in that last year. I had to, in order to make sense of my narrative—figuring out the narrative and lessons learned, and also revising as I went, and as the story moved forward. I could read a novel now—was this the end? I could exercise with vigor now—was this the end?

People ask me now if I feel 100 percent recovered.

For a while, I said I'd never feel 100 percent again, even though I knew the point at which I felt I would be okay would keep advancing.

I was driving home one day from wherever I had been out—I don't remember where. That part of the story does not matter. It was in the late fall of 2008. There was traffic on the freeway. That part of the story matters. Because when I am stuck in traffic, I basically had a habit of staring at license plates. And I realized that I was memorizing license plates to while away the time.

I was memorizing license plates. Just as I did before the stroke.

I knew then that I was okay.

I knew then that I was well.

But was I better?

When people ask how you are, and you are healing, you're supposed to say, "I'm better." I did not want to say I was better, because I thought I would never be better. And

I was sick of making people uncomfortable, so I just said, "I'm still alive." Which was true.

How is it that "I'm better" so often means recovered from an illness?

Better means improvement.

Does that mean you are better than before? Or just better than the month prior? Do you say "I'm well"? What is wellness?

And how hard it was for me to have lost my old superpowers and recognize new strengths.

Was I better? Yes. Better than before?

The thing is, I am better. I am better than before.

I have a different answer to the question about whether or not I'm at 100 percent.

These days, I answer that I am. I am a different 100 percent. I tire easily and I still have occasional memory problems. But in the place of these shortcomings, I have new wisdom and resilience.

My story did not end at my recovery. It continued, with ramifications far and wide.

Being sick and being in recovery gave me a tenacity that I didn't have before. And that tenacity has been key to my life as a writer. Never give up never give up. The scars sickness gave me also gave me a soul as a writer. This is not to say I didn't have scars already, but these particular scars tore at my ability, and tore at my intellect, the pillar of my self-esteem and identity. I was wounded in a way I'd never been before. I was disarmed. And because I was disarmed, I had to live life in a new way. I became different. At first I resented this difference. And then I accepted it. And then I loved it.

No one told me that getting better meant becoming different. And that different meant a sense of loss. And that the loss would eclipse improvement until I saw improvement.

And that I would be so so so restless for an entire year. For years afterward. Restless with myself. Restless with life. I traveled. I would move to New York City. I would dream of being anywhere but where I was.

Five years after my stroke, when I started teaching community college, I made it a point to remember my students' names within two weeks. Whereas before I would have just memorized the names, now I had to write notes by each name, things to remember students by. Memorization became a conscious act. Explicit.

And yet, my memory itself still relied heavily on implicit memory, in which previous experiences aid the performance of a task without conscious awareness of these previous experiences: riding a bike, driving a car, dialing a phone, typing on a keyboard.

Writing fiction is a mix of explicit, implicit, and emotional memory.

The characters and plot are explicit.

And then they are implicit.

And then beneath that is emotional memory.

The act of writing is implicit.

The act of composing sentences is implicit.

The act of imagining a world is explicit.

The goal is to write a story that enters emotional memory. To make it so readers will remember your story and your writing, because you have struck something in their heart until it reverberates in their mind.

15

I picked up *Slaughterhouse-Five* a year after my stroke.

I read the book in its entirety. I learned that the protagonist was not Vonnegut, who speaks directly to the reader in the first pages, but Billy Pilgrim, who appears later, in chapter 2, the first line of which is "Billy Pilgrim has come unstuck in time." Oh.

The protagonist in my own story also changed.

———

Trauma takes time to stare in the face. It was difficult to examine the pivotal thing that injured me, disturbed me, and changed me forever. Looking back at my stroke and the recovery made me remember having to hold people's hands through airports because if I got lost, I wouldn't remember where to go. It made me remember looking at words and seeing them as shapes and not knowing what they meant. It made me remember the friendships I lost because in depression I was unkind. I am still capable of being enormously unkind, but in those days I had no restraint in my unkindness. I was unable to keep my insults inside my mind. Instead of telling people they'd let me down, I called them names. Instead of trying to understand why someone was slow in line, I immediately raged. It made me remember the day Adam tore up the carpet, and how the rug had been pulled out from under my own life. It made me remember how no matter what you do or how much you follow the rules, something bad could still happen to you. And in the wake of it, you could be reduced and helpless.

It made me remember, because now I could.

Trauma is never forgotten. When I see a new mother holding her infant and going through the repetition of feeding, burping, changing diaper, napping, my body remembers the ceaseless cycle of new motherhood. My lungs constrict, my heart rate increases. The sound of a breast pump makes my muscles flex, makes me want to run from the room, even as I plant a smile on my face, hoping no one notices my panic. Long, unexplained absences by my partners will always remind me of Adam's inexplicable trips away from home the year our daughter was born, and I will lay awake at night, my fingernails carving half-moons into my palms. Every headache will require a double check to make sure I am not stroking out.

My mind will make up stories out of fear—will tell me I am never to be loved, that I am incapable. Even while the mind knows that I am lovable and competent and there are extenuating circumstances for each life surprise.

Remembering trauma is to feel that trauma all over again. I thought I had to have distance to see the thing as a whole, and so I tried my best to divorce myself from the event and to put it behind me—but I learned that trauma is not a separate thing, it is an experience that folded itself into my body and mind and brain. I had to hold it close, sit with the discomfort and pain, and not push it away. It took a while to learn this. And when I did, it felt like holding a turd in my hands. I held the fear and hurt and pain and confusion, long desiccated. I didn't know what to do with it. I stared at it. I held it at arm's length. It was disturbing. And then it was awkward. And then I became used to it. And then it became valuable.

Historically my way of dealing with disturbing things was not healthy. "Put all your uncomfortable feelings and memories in a box," I often said, "and then put that box in the darkest, dustiest part of the basement of your mind, where the spiders live." And then I would laugh. "It's in a box—that counts as compartmentalization, right? Even though some people call that repression?"

But that is not a sustainable strategy. The thing that I had hidden in the recesses of my mind is the thing that found its way out in everyday dealings.

The ignition sound of a Honda will make my heart jump, because my dad followed me to school every day and I knew he did but he always said he did not and only twenty years later will he admit that he was following me on my walk to elementary school each morning because he wanted me to be safe and he had to make sure even though it actually made me feel like I was being stalked. My dad did this because his world was never safe, because children died all the time in his

world and he felt helpless then and did not want to feel helpless with his own daughter. In his experience, the world has
fallen apart. In his experience, families get separated. This is
his trauma. When I said following me around was creepy, he
said he wanted to protect me, even from a distance. He did not
mean for me to become haunted.

The triggers are many.

When I get a B+ on my report card I will feel like the world
has ended because my parents tell me that I must define
school and academics as life or death. That I must study as if
my life depended on it. And that anything less than an A will
get me killed. Because that was the truth for them, at least in
1960s Korea, where jobs were scarce and resources few. And
this train of thought saved them, kept them going, and they
want more than anything for me to be strong and disciplined,
and they are sure they are guiding me along the path with the
most guarantees. And even though it is no longer the 1960s
and this is the United States and I live in the suburbs, I have
lived with this stark black and white view my entire life to
date and my parents trust it and so do I, so I do feel like the
world has ended, and I wonder how to move on from here,
and I wonder how I could have ever let it get to this. I wonder
why life feels so wrong and horrible and heavy and I am fifteen years old, and I feel like I have been through a war, even
though I have had every comfort in life. And it is because I
have been through a war, because my parents went through a
war, because my parents were able to climb out through their
education and that was their salvation, and I do not really understand this, but I feel it. I feel every ounce of it.

I will start cutting, because I have learned to compartmentalize everything. I have learned to dissociate myself from emotional pain and sadness and grief, because those things do not belong anywhere in a war. People who feel sad during wartime become dead people. The only acceptable feelings are anger and happiness, and I become a person whose depression is overlooked, because I am an angry person now. I will not know what to do with the other feelings, and they will come up. Because they always come up. And I will try hard to numb myself, until I can no longer contain the sadness and self-hate—so I cut myself to release what I no longer know how to release. I take a scalpel because my mother is a nurse and I have found a scalpel in her pocket and I have snuck that scalpel away for myself and I slice my skin so that I feel physical pain, which is a relief. It is a relief. I do it again. And again. And again.

I still wince and crouch as I cross streets, years after being hit by a car in Seattle. I will never cross a street while a crosswalk light is counting down—not after a car hit me while I was in that ambiguous space. Because one day I was in the crosswalk, and the number was counting down from ten and there was the sound of an engine revving, and as I pivoted, I saw an SUV grill hurtling toward me right before I heard the sound of brakes and before I heard the sound of garbage cans being hit before I realized that was the sound of the car hitting my chest and cracking my sternum before I realized I'd fallen first on my right thigh before my body twisted and flew and before I finally skidded on my belly and I would have a dark purple bruise for three years and I will never walk in a crosswalk under those circumstances again. Ever.

I will be watching *The Girl with the Dragon Tattoo* on the

television, and I will sit through the scene in which Lisbeth Salander is raped. After, she will bleed in the shower, and I remember that after my rape, I showered and used Dr. Bronner's peppermint soap, which stung the parts of me that bled, and I will tamp this vulnerability down into the deepest part of my psyche. This did not happen, I will tell myself. This did not happen. I trust this man. He is my boyfriend. It was an anomaly. He must not have heard me. I must have done something wrong. I was punished. I deserve this.

Years later I will meet up with that same man to tell him I wrote an essay about him. I do not ask him for his permission. He is married with children and we have remained friendly through the years. He will say sorry. He will divulge his faults. He will say he was an alcoholic then. I will say we were different people then. I will quail when we hug.

I will wait and wait and wait for my husband to come home, but he does not. He says he is at work. He says he is on a business trip and it has been extended. I wait with my infant daughter in my arms. I am angry. Because I am depressed. I am filled with rage. I am turning that rage inward on myself. But no one knows I am depressed just yet; not even I know it. I have not yet realized that this is postpartum depression. I will not realize it for months and months. I will never ever want to wait for a man to come home, ever again. And when I do, I will be overcome with huge panic and anger and it will make me want to leave, instead of being left.

So instead, I organize the baby bottles in the dishwasher. If I have lost control over my entire life, if my life as I know it is no longer, then at least the baby bottles can be arranged in perfect order, and they can be sterilized. All the bacteria and all the germs can be killed. All the invaders must be excised. Something, anything, has to be perfect. Something, anything, has to be the way it was, back to the way it was before. I fill the dishwasher, empty the dishwasher, pump my breasts, fill the bottles, feed the baby, and then I arrange the bottles and sterilize and sterilize and feed and sterilize and feed again and again and again.

When Adam does come home, when after all the truth comes out, he comes home through the door with his head low, and I hug him. It is, in hindsight, a hug good-bye. We cry. And then I cry. And cry. I cry for weeks on end. I feel sadness. And because I allow myself to feel the very emotion of which I have been most afraid for my entire life, the anger does not come. Instead, I heal.

I am disappointed. And then I accept. Then I understand.

After Adam leaves me and later I fall in love with another man and that man, Mr. Paddington, snaps his fingers while recording music, I will automatically respond with, "What?"

And my boyfriend, Mr. Paddington, who is also a songwriter who is also a software engineer who is also an English major from UC Berkeley who graduated with me in the same class who I have known for years who understands dying and darkness and light who was Adam's close friend will say, "What?"

And I will say, "You snapped your fingers. What do you need?"

And he will say, "I was just snapping my fingers to make sure the mic was working."

And I will realize then that my ex-husband used to snap his fingers to summon my attention. I will realize it started as an innocent joke, but it did not end that way. And I will try my best to not twitch when my boyfriend snaps his fingers again and again while he records music over the years. And it will take two years of finger snapping until I do not have to suppress the urge to leap up and say, "What?"

PTSD, or post-traumatic stress disorder, is a result of avoidance.

I thought that the only way I could see my stroke was to distance myself so that I could see it in its entirety. So I detached from the event. It was, I thought, something that had happened. I put my stroke into a box in a basement. I wanted it to stay there.

Time to, I thought, move forward.

So I did. I went back to work. I went back to school. I tried hard not to talk about my stroke. I avoided the online stroke bulletin boards, on which I relied so heavily in the first year of recovery. The topics felt irrelevant to me. The members felt so injured and damaged, I wanted nothing to do with them. When stroke survivors wanted to discuss their stroke with me, I cut the conversation short. I did not bring up having had a stroke with new people in my life. In fact, even now people are shocked when I say I am writing a memoir, and that the memoir is about surviving a stroke.

I wanted everything like it was before. I tried not to think about it. It had happened. I thought it was no longer happening.

In avoiding the thing, I could not avoid it.

I told myself to not think of the stroke.

I felt no one wanted to talk about it—not my cadre of healthy friends, anyway. When I said, "I can't remember— ever since the stroke, I can't remember little things," they told me they could not remember things, either. That it was old age, that it was their recreational marijuana use, that it was exhaustion, that it was normal.

Not the stroke.

I fell down the stairs. Occasionally, my mind would burp. It still does. I have to then tell myself left foot right foot left foot step step step. Do not fall down the stairs. Do not fall down the stairs. In telling myself to not fall, I would fall.

Not the stroke.

I was an extrovert. And now, I was an extrovert in an introvert's body. I craved social interaction, but socializing exhausted me. At conferences, I would make sure to book a hotel on the premises and take frequent breaks. "I need introvert time," I would say. And then I would dash upstairs to my room to sit in a chair, stare at the wall, rest for an hour or two.

Solitude was replenishing, like a long drink of cold water on a hot day. I needed it. I'd never needed it before.

A new friend, who had only known me as extremely social, asked another friend, "Where did Christine go?"

"She went upstairs! She said she needed down time."

"Wow," said the new friend, "I thought she was kidding!"

Not the stroke.

I mix my homophones. Just a few paragraphs earlier, I typed "stair at the wall," before I caught myself and retyped, "stare at the wall."

Not the stroke.

Adam and I resumed our life. As best we could. So much had happened. His mother had died. I had been sick. He had a start-up to run. Everything was back to normal. Everything had to be back to normal.

Not the stroke.

Not the stroke.

Not the stroke.

Until all I could think was Not The Stroke. So that the stroke was the main focus. I am sleeping because I am tired. Because of old age. Put it behind me. It never happened.

Not the stroke. Not the stroke. Travel to Europe. Travel to Beijing. Go to school. Make dinner. Not the stroke.

It took me years to be ready to write about my stroke. I did try to write about it. I wanted to write about it. I attempted to write an essay about my stroke no fewer than eleven times—there was no narrative, there was no structure to it. I wasn't ready. All in all, it took eight years to write it down. It took that long because I tried for eight years to put the stroke behind me. It wasn't until I had a baby and I had postpartum depression and my husband of fifteen years left me that I could look at the stroke.

I looked at the stroke because it was easier to look at that earlier trauma than the one before me. Each time I have overcome a trauma, it was by facing, in some way, the trauma preceding it. But this book—this book made me look at everything, and it has made me come to terms with the things that happened. It made me hold everything close. My daughter.

My memories. My marriage. My mind. My brain. My body. The life before. The life after. What they mean. What they meant. It made me sit with my entire life. This book is about my stroke, but the stroke helped me come to terms with other traumas, including the biggest trauma to date, the end of my marriage to Adam, my college sweetheart, my daughter's father, and the man who had been the center of my world.

Writing trauma feels impossible at first. Vonnegut, in the opening chapters of *Slaughterhouse-Five*, confesses as much. He created a character to embody his trauma and exorcised the darkness in that way—Billy Pilgrim traveled through time. He knew how and when he would die. The defining moment of his life, the only time the narration moves in chronological order, was when he was a prisoner of war during World War II in Dresden. Vonnegut wrote this. Kurt Vonnegut died on April 11, 2007. Not long after my stroke. And while I was unstuck in time myself.

I too have created a character to embody the trauma. The person on these pages is the character Christine. The person telling you this story is the narrator Christine. And the person behind all this is me, Christine.

That was I. That was me. That was the author of this book.

The character and the narrator and the author all changed. And maybe the reader changes, too.

My stroke is not the only trauma in my life. But it is one I can share freely, the thing I feel I understand, the thing from which I have learned, and the thing that is not who I am, but a part of who I have become. I used what I learned from other traumas to help navigate my stroke recovery. And I have used what I learned from my stroke recovery to maneuver new tragedy.

Trauma can only be seen once held close.

To overcome trauma one must confront the event. Over and over, even. I must say to myself, "All this happened."

And as in *Slaughterhouse-Five*, once I get to the end, the end is the beginning is the middle is the end is the beginning. Time keeps looping. Memories return. Trauma, in the end, is the beginning. It keeps evolving. The idea must be revisited. Over and over again.

And with the idea come the fears, which the mind makes up, which is the psychological manifestation of physical trauma. I had anxiety attacks. I had high blood pressure. I was in a very stressed state, as if under attack, as if immersed in war. In looking at my stroke up close, I processed the fears I'd had. In looking at the stroke, I examined the commonalities between traumas.

When I had a baby, my entire life changed—I lost abilities and purpose and identity, just as I had after my stroke. I was also exhausted. The last time I'd felt so exhausted, I told people, was right after my stroke. All I wanted to do was sleep, but I could not.

When I waited for my husband to come home to be with me and our daughter, to help, to make things better as he always had, it reminded me of waiting to be healed. Of waiting to be myself again. Of the stroke recovery process. Of the helplessness.

When our marriage ended—when he and I reconciled that we had become different people—too different to live our lives out together, it reminded me of the way in which I had to say good-bye to my old self in the wake of the stroke. Adam could not have his mother back. I could not have Adam back. The ways in which we could come back together no longer worked. Sometimes partners lean in, as Adam and I had during so many challenges during our eighteen-year relationship. But in the end, the stroke and his mother's death proved to be too much. I changed. And so did he. We did not know this until years later.

And in the end, Adam and I broke each other's hearts, albeit at different points.

Marriages can become stronger in the wake of tragedy. Ours did not. Could not. Would not.

A deep grief and sadness poured through me. Who was I, now, without my lifelong partner? Who could I become?

I was on the treadmill a few days after LASIK so I could finally see again without visual aid for the first time in thirty-four years. The stroke was more than eight years in the past. One eye was still a little blurry—and I was okay with that, because I could see well enough to drive. And day by day, my brain was getting used to the blurriness. The brain was compensating. The brain was plastic.

And I was running. But waiting for the pain to hit.

Because I always wait for the pain—that dizzy feeling of being choked to death accompanied by a visual aura—to hit when I run. Even though it now never does. That is the way trauma manifests. Although since my PFO closure I no longer experience pain every time I run, my mind still thinks it will happen again.

The infarct's effects are largely gone, but it is also still there. I will always have a history of stroke. A part of my brain will be dead forever. But there was also a miracle. The other parts of my brain compensated for what was lost. The right side of my brain took over the functions for the left. I was reorganized, just as if a living room in a house is destroyed, the playroom takes over its functions. The previous rules no longer apply. Everything that happens to us in our life is present, forever.

———

A red snowblower reminds me of winters in Tahoe—how the neighbors all came out of their houses on weekend mornings to blow the snow out of their driveways, like lawn mowers in the midsummer suburbs. How we all waved to one another as we pushed the snowblowers up the driveways, down the driveways, and then up again. Red snowblowers are more than these weekends; they are the image I associate with a dying brain in wintertime. How they looked like cherries in the snow. How they were lined up in front of the store. How everything tilted. How everything turned nearly upside down.

Getting lost no longer petrifies me. I used to map out the entirety of a trip, and never wanted to end up somewhere unplanned. And then that map was taken away from me. That map was no longer accurate. I had to roam lost. Maps now have different meaning; they are guides. They are an illustration of one way to a destination. Being lost has its merits.

I no longer hold grudges, because I cannot. Resentment

is too exhausting and I no longer have the energy to sustain bitterness.

Emotions and words are a larger part of my life than they ever were.

My wiener dogs are now dead. I still squeal when I see a dachshund in passing. I always associate them with Fritos, because that is how they smelled when they needed a bath.

A Frito pie will always remind me of that small kiosk in Austin, Texas, and a hot summer evening before celebrating my friend's wedding.

I loved weddings. I loved our wedding day. And I still cling to hope for the couple, I cling to the wedded moment when late nights alone and waiting waiting waiting are nowhere in their minds, nowhere in their lives.

Waiting is an accepted part of life. I waited to recover. I tried so hard to push time and push myself, but in the end, things unfolded. I had control over only so much of the outcome, and I had to learn to be flexible and open and porous to the lessons and pain and sadness and hurt in order to survive and then thrive.

My dream was writing, and after my stroke I wanted to write again, and every word I write now is a victory. I hid my writing as I hid my true feelings, and now I cannot waste time hiding the very thing I want to do most. There is no more time to waste.

———

Memory never really leaves. I was unable to retrieve my memories in the immediate wake of my stroke, but they are all somewhere in my brain and in my body, informing me, becoming part of my intuition, if my mind cannot remember. Memory is part of my body—a baby crying makes me wince, makes me rock my body in a motion of comfort. When I am distraught, I will hug my knees and rock, because somewhere deep inside of me, my body remembers its first method of comfort.

Part of trauma is about being stuck in the past.

According to Bessel van der Kolk, war survivors who undertake Rorschach tests see every inkblot as a vestige of their psychic wounds. In the abstract shapes they see a dead baby, a dying friend, or nothing at all—they cannot see beyond their trauma. "War," says van der Kolk, "can break the projector of imagination, and the only option is to play the same reel over and over again, or turn the machine off altogether."

For years I played my stroke over and over again. It happened, it happened, it happened. How did it happen? This is the way it happened. I was in a parking lot. It had snowed. I had a headache. Everything I saw reminded me of those snowblowers. Everything I felt reminded me of how my world had turned upside down, just as the snowblowers had turned sideways.

I want to know what it all means to me. The headache and my brain cells dying. How my confusion and lack of speech and visual disturbance were the results of the clot blocking blood flow and killing my thalamus and making it so that my brain was unable to relay information as it had. I want to know the stroke's lasting implications in my life. The stroke's role in my marriage. The stroke's influence on my writing.

I thought of the stroke then. Now the stroke folds into my life.

The stroke's effect has never ended. It ended my marriage.

We have all moved on. And we keep moving forward as life's calamities befall us.

There is a myth that the Amish deliberately include a mistake in each of their quilts. And the flaws make the quilt more beautiful, more individual, and unique. It is called a Humility Block. Whether or not this myth is real, the idea is true.

I visited the stroke unit at John Muir eight years later. I wondered if the rooms and the beds and the nurses and the smell of disinfectant and the sound of the chirping telemetry machines would be the same as I'd remembered. When I got to the second floor, I followed the signs to the PCU, the adhesive nametag on my blouse stiff against the pliable cotton, my arms stretched out holding a box of croissants and morning buns and pain du chocolat. I felt like a visitor. I was a visitor. And then I got to the last bend, at which point everything felt familiar, like visiting a childhood home, like riding the 7 train lost in reverie and then realizing the sound of the subway train on the elevated tracks was the sound of my childhood, like comfort and safety and intimacy.

And there were the rooms. They were there eight years ago, and they are there now. There are patients in the rooms. There were patients in the rooms then. They all eventually leave with memories of the rooms and the ward and the nurses and the smells and the sounds. It goes on and on. What happened keeps happening.

There was my room, the first on the right.

"Hi," I said, placing pastries at the nurses' station. "I was a patient here once."

"We don't get many return visitors here! No one ever comes back—to visit," said the head nurse.

She had the dark humor that I've come to associate with nurses.

I nodded. "I suspect no one wants to come back. Or—they can't."

I told them I was writing a book about my experience. I told them I wanted to come back and get a feel for my time there, for the things I'm not sure I remembered correctly. To see what had changed over time. To see what memories might be shaken loose by visiting.

They asked me what it was I wanted to see.

I asked where my favorite nurse, Fumi, was—was she on shift? No, she was not. I vowed to return to give her a hug. Most of the nurses on staff eight years ago were still on staff now. They stared at me and I stared back, as if at something from another dimension. And it was—it was from another time and space. Here we were, all of us healthy.

I wanted to see the shower. The tiny shower.

I inhaled the smell of the progressive care unit.

I cocked my ears to hear the beeping and chirping of the monitors.

And then there were the doors, the double doors demarcating stroke survivors from patients with other ailments. You are one thing before entering. You are another, afterward.

Eight years later, my stroke feels like the beginning of what I now see as a turning point in my life. This is not to say that the stroke was the best thing that happened to me— because it wasn't. I could have gone my whole life without having had one, and that would have been more than fine with me. But it turned the boat of my life around.

My life had been lived according to a script, for the most part. And I thought I was writing the script. But I am not the author of my life. I do not control what happens to me. I am the narrator, the protagonist who has learned more than one way to live a life. And that there is more than one way back from a setback.

Adam and I are friends. And that is the outcome of the lessons gleaned from my stroke. We made it through the stroke together and emerged as different people. And I made it through the stroke, and while processing fear and letting go of control and embracing sadness, I learned to accept that bad events do not have to remain bad events.

Billy Pilgrim traveled through time. The only parts of *Slaughterhouse-Five* in chronological order are the parts set in Dresden. Otherwise, the novel flits back and forth between past and future with a connecting theme of war and postwar.

A story is told in modules. That is the structure of story.

When my memory began to come back, I became acutely aware that I remembered only certain parts. I would trail off, until someone reminded me what happened next. And even then, the story as that person remembered it would not be mine. There were things going on that no one else could see. And I have done my best to write them down here. To weave them all together in a way that makes sense.

There are notes in my journal that told me what I thought and did each day. That I brushed my teeth and then was tired. That in the beginning I did not understand what was going on around me. That I traveled tons.

But it is my mind today that had to make sense of those events. The links between my travel and my emotional state. My avoidance. The lessons, like how waiting is part of the healing process. That I developed new resilience. That I developed tenacity. That that tenacity made me a better writer.

I wrote a novel for ten years—during which time I had the stroke and then a baby. I did not think 1 would ever be finished. I had written 110,000 words and thrown them all away. I started my novel again. I wrote another 70,000 words. I threw them all away. And then I rewrote the novel again. And again. And again.

The language of my stroke is forever in my brain.

My stroke laid bare the inner mechanisms of my brain, the way words are formed and then connected and the way ideas link into a story. It is only by losing the ability that you can truly understand its place; stories are told in modules. One thing leads to another.

The stroke happened.

And then the hole in my heart was closed.

And then because the hole in my heart was closed, I could exercise.

And because I could exercise, my mind forgave my body.

And then because I was whole, I became pregnant after thirteen years of infertility.

I had a baby.

My mind fell apart. My marriage fell apart. My heart broke.

And in that heartbreak, I wrote an essay. I wrote about the thing that had hurt me most but was not the thing that hurt me most at that time. I wrote about my stroke, which was so much about my divorce, but not about my divorce, but all the lessons I learned to survive my divorce. I wrote about my stroke, which was the epicenter of this earthquake.

My baby was in my arms and she was growing up and all we did was sleep and eat and burp and shit and piss. Through the night. In two-hour cycles. We all knew my marriage was ending.

I feared being alone, until my friend Chelsea took me aside and said in the gentlest of voices, "Christine, I know you think you can't do this by yourself. But you've been doing it by yourself all year."

My life fell apart and then it rebuilt. Everything healed. And life started again.

It would all begin again.

And again.

And again.

16

1.

Listen—it is 2013. I am back at the same hospital. I am here again, for a completely different reason than the stroke, but that is related to the stroke. My healed body is here to birth a baby.

I have stayed in the labor and delivery unit before, for a completely different reason that is also a related reason. The first night I was admitted after my stroke, there had been no beds in the progressive care unit with all the other stroke survivors. I was given a bed in L&D instead, my needs a subset of what pregnant and laboring women require: a bed, monitors. The hallways are familiar to me. The gown is the same. There is an eerie déjà vu. I time travel.

I am here to give birth, I tell myself. I have not just had a stroke. The walls are the same walls, but I am not the same person. The pain is not the same pain. The walls do remind me of the old person. The new pain does remind me of the old pain. My rising blood pressure, which necessitated the scheduling of this induction one week past my due date, does remind me of the stroke.

Adam and I make ourselves comfortable. We set down his bag of snacks. There are granola bars and microwave meals and chips and cheeses and bagels and cream cheese and pea-nut butter and apples and oranges and juice boxes in there. Enough for a week, if need be. Later, our doula will tell us upon seeing that bag, "This baby will be here soon, then. The more you prepare for a longer stay, the faster the baby comes." We call our doula, Felicia, to tell her we are here. We are about to begin. It is time.

I am hooked up to monitors. I have a blood pressure cuff. This, too, brings up memories.

The Pitocin drips. There is a high-decibel beep that accom-panies the moment. Precisely twenty seconds after the Pitocin releases, my body reacts: it feels like dysentery cramps along-side constipation. For two years into the future my body will brace itself after hearing beeps. I do not scream. I breathe. Scientology would be proud of me.

The nurse comes in throughout the day to note the time and to check my progress, which is to say how many centime-ters I am dilated. 9:00 A.M.: 4 centimeters.

At noon I get my scheduled epidural.

"I still want to feel," I say to the anesthesiologist.

"Why?" he asks.

"I want to feel everything, but just a little bit. I want to move my feet, even if I cannot walk."

"Are you sure?" he asks.

"I am sure," I say.

When he is done, I can move my feet, even if I cannot walk. He has done his job. He has given me a good experience.

Adam glances at the monitor. "Wow, that was a huge contraction."

I blink. "Didn't feel it at all." It is an immense relief.

Felicia tells me, "Take a nap now. Get some rest."

And I do. I sleep.

2:00 P.M.: 8 centimeters.

The doctor comes by during rounds, breaks my water to speed things up. They also turn the Pitocin up.

Two hours later, I am fully dilated. 5:30 P.M.: 10 centimeters. It is time to push.

2.

I am pushing my daughter into the world. Everything that day is counted. Induction is to begin at 6:00 A.M. The amount of Pitocin is precisely measured. The level of contractions is monitored on a screen. The minutes between contractions are recorded. The seconds of pushing are chanted. Push for ten seconds. 12345678910. Push. Bear down. Felicia tells me I am almost there. She says my daughter is on her way.

I told Felicia the one thing I needed during delivery was for her to tell me how much time there was to go. I told her that I was used to pain—but it would help to know for how long that pain would last, so I could break it down in my mind. Two days of pain is different from one day of pain is different from six hours of pain is different from one hour of pain. To know the end, something I did not have for my stroke recovery, is reassuring. In this way, I have learned from my pain.

The doctor is called. She too is on her way. Everyone is on the way.

There is not much time remaining.

3.

Hold. Now time to breathe for ten seconds. Deep breaths.

Inhale. Fill my body up with all the air, all life, all strength, all flexibility. All light. Draw in the moment I found out I was pregnant, alone in the bathroom, holding a pregnancy test with a shaking hand. Breathe in all my trips around the world breathe in love and falling in love and holding Adam's hand in college and holding Adam's hand in the labor room breathe in happiness and joy and weddings and sunlight through veils and breathe in the last word of a finished manuscript. My body floats. My body buoys.

Exhale. Breathe out pain breathe out darkness breathe out fear breathe out exhaustion breathe out heartache and waiting and unrequited love and breathe out inadequacy and dreams deferred. Breathe out the past and future and focus on the present and breathe out my old life.

Inhale again. Draw in warmth. Take in horizons take in full moons take in cool breezes on warm days. Think of yoga. Think of wide spaces. Think of blankets. Think of flying. Think of my body expanding. Think of bending.

Exhale. Breathe out heat breathe out ice. Breathe out pain. Think of purging. Think of clean clothing. Think of lying down.

4.

Adam holds my hand.

Eighteen years we have been together. Thirteen of those trying to have a baby.

He holds my hand. He will be meeting his daughter, too.

5.

The machine beeps.

The room is warm from the lights. We all speak in hushed voices, as if this is a holy moment, as if we are in a yoga studio as if we are in synagogue as if we are in a library as if we are in a lecture hall as if we are in a museum.

My legs are warm from the hands holding them.

"Can you turn the epidural up?" I ask. It's getting unbearable.

"I'm sorry, Christine. This is the part of the labor pain the epidural doesn't block," says Felicia.

"Oh."

6.

The doctor arrives. "Hold off! I'm not finished scrubbing in yet!"

I look over to see her garbed in scrubs, brushing away at her hands with surgical cleanser. The swift, firm strokes are frantic. I am splayed on the table, my legs up in the air, my modesty long gone.

"Look at how perfect the timing is," I say. "You'll be home in time for dinner!"

She laughs. Her laughter is louder than her scrubbing. Better.

7.

Now push for ten seconds. Then push again for ten seconds, no break.

Push hard. Bear down. Push until all the oxygen leaves my body, until the capillaries on my face break, until I feel I have pushed everything out of my body, all the years and all the heartbreak and all the illness and all the waiting. Out.

Inhale quick. The whole world in my lungs in one deep breath. Lift. Bend.

Push again. Exhale everything. Everything. Follow my body as it deflates.

Inhale. Push.

There is no rest. There is the work.

8.

My daughter breathes air. She, too, currently has a patent foramen ovale, but it will close as she grows. She, too, has had an exhausting day. We have made this journey together—emerging from a body is difficult work.

The doctor notes the precise time.

That time will be the time of her birth, forever and ever. Today is her birthday.

9.

I do not have the mythical moment mothers feel when locking eyes with their newborn. I do not fall in love at first sight. I will fall into a deep postpartum depression out of which I will claw. But eventually, I will fall in love. I will fall in love again and again and again. With her, with life, with writing, with friends.

The birth of my daughter will be the best thing that year. The best person that ever happened to me. That moment preceded by all the moments previous. That moment, impossible without struggle.

10.

There is space in my brain. There is space in my body. There is space in my mind.

My body is no longer at war.

ACKNOWLEDGMENTS

On February 14, 2017, *Tell Me Everything You Don't Remember* will be published and out in the world.

I did not do this alone.

In 2013, I lost everything I'd built my life upon. My marriage. My previous identity. Money. I was heartbroken and dealing with postpartum depression. I was struggling with motherhood and the challenges of my new life.

I began a relationship with my newborn daughter that same year and I found my identity and strength and friends and love. Everything familiar was gone, but I had the opportunity to replenish my life with the things and people most important to me as a newly untethered individual.

I remember telling Mr. Paddington that I had one year to really make a change—that for a year I would be at home as a new mother, I would have no money, and that that would be the year I would double down on dreams. Everything I know is gone, I told him. I have nothing else left to lose. I have to do only the things I love to do and see where they lead me.

I felt helpless, and so I did the one thing that did not make me feel helpless: I wrote.

In 2013, I wrote an essay, *Mint,* which in hindsight was a turning point in my career. Published in the Rumpus by Roxane Gay, this piece changed my life. I thank Roxane and everyone who read *Mint.*

That essay led to an opportunity to write something for BuzzFeed's longform section, Big Stories, in 2014. I wrote an essay about my stroke and recovery. The essay went viral and led to a two-book deal with Ecco.

Thank you to the great Isaac Fitzgerald, who made me known to BuzzFeed. And thank you to Cheryl Strayed, for introducing me to Isaac in the first place over a Thai dinner in San Francisco. Deep gratitude to Sandra Allen, whose editorial hand helped me to write an essay of which I am proud. And to Lisa Perrin, whose pitch-perfect artwork graces both the cover of this book and the original BuzzFeed essay. Thank you to all of the people who read and shared the essay, like Mark Armstrong of Longreads, which made it go viral, which brought it to the eyes of agents and editors, which brought forth this memoir.

My acquiring editor, Hilary Redmon, helped me get this book on its feet; her guidance was very much a part of the subsequent first draft of the memoir.

Megan Lynch and Eleanor Kriseman made sure we were in lockstep throughout the entire gestation of this novel. I know this book is better for having had their eyes on my words. I am excited to kick off my career with them.

The generosity, ferocious spirit, and knowledge of my agent, Ayesha Pande, kept me feeling both safe and elated throughout. And still does.

The year 2013 was an enormous fall. On an autumn day

that year, Mr. Paddington and I took time to coast down the concrete slide in Berkeley's Codornices Park. On that day, I decided that as miserable as I felt I would seek a minute of pure joy somehow. My thinking was that I could hold on to those few seconds and say, "Today I felt good, even if for ten seconds."

That is how I clawed my way back. I would hold on to the small parts of good, even if the good parts were just one percent of my day. I would then try to expand that one percent however I could. I would hold on to any part of happiness, even if fleeting.

My friends helped me to enable those little sparks of joy and reached out to me when the world felt like a dark well.

Thank you to my friends, who always returned texts and sometimes returned them as phone calls. From the day I was distraught and rocking myself on my bed to the day I signed with an agent.

Randa Jarrar, for your strength—we don't lean in; we eat that shit for breakfast. Connie Wong, for your stalwart friendship. Ruth Halpern, for steadiness. Lizz Huerta, for your sass and a voice on the other end of the phone always. Karissa Chen and Eugenia Leigh, for your deep loyalty and example of friendship. Patricia Engel, for your wisdom and loyalty. Alexander Chee, for your brotherly advice and guidance. Nova Ren Suma, for your dedication. Ankur Parikh, for telling me the truth. Candice Carr, for our adventures and laughter. Wah-Ming Chang, for your generosity. Laura and Peter Fenton, for your steadfast support. Cathy Chung, for your clarity. Krys Lee, for never being afraid of the darkness. Ruth Saxton, for all the conversations over tea that we both will never remember. Rose Mark, for your tenderness. Aimee Phan, for your kindness and for cookies. Matt Salesses, for your vigilance. My fairy godmother, Sharon Plonsker, for your deep compassion. Tony and Terry Letizi, for opening up your home to me and Penelope. To my doorpeople in New York City, Luis Marte and Daphne Loiseau, for showing me hospitality long after I moved out of the building. Adriana Hartley, for showing up. Naomi Williams, for coming over to visit me when I was alone. Nina LaCour and Kristyn Stroble, for your display of amazing partnership and for my first mother's day gift. Amy Shin, for helping me without wounding my pride. Andrew Whitacre, for your camaraderie in the wake of my stroke. Heather McDonald, for your ferocity as a writer. Reese Kwon, for your resourcefulness. And my dear departed friend Tim Eck, who once cradled me to sleep as I cried.

Thank you to all the readers of my anonymous blog, jadepark.wordpress.com, at which I wrote about my recovery before I wrote an essay about my stroke and long before I wrote this book. On that blog, I heard from old friends who cheered me on and new-at-the-time friends who have since become lifelong friends and supporters. The blog was a key place for my recovery—a place of safety and low-stakes writing where I could sing out my words, even if I stumbled.

My friendships sustained me, and 2013 was the year I discovered that family goes beyond blood. I would focus on happiness. I would be aware of misery and I would try to deal with the bills and legal paperwork one by one. My worries were many—at one point I wondered how I would pay for diapers. I would not ignore these concerns. But I would look at slivers of happiness while dealing with the unpleasant.

And, eventually, the happiness would dominate.

And, yes, it has.

Throughout 2015, I wrote my memoir. I am grateful that my daughter was under great reliable care so that I could focus on writing. Thank you to my fleet of caregivers: Chelsea Johnson, who showed me that there are people in the world built of kindness and a spirit of altruism; Juliette Mueller, for your clear-headedness; and Amanda Hsieh, for your sense of adventure. To all my doulas: Felicia Roche, Pat Mullarkey, Peggy Hinkle, and Carmen Fraser, for your immense care. Also thank you to the Mulberry School.

I wrote my memoir, I did research, and I did my best to verify facts, but I also had support in doing so.

Special thank you to Dr. Dagan Coppock, who read this book for medical accuracy. I met you at Squaw Valley, and I am grateful we each had the bravery to say hello to one another and begin what has turned out to be a lifelong friendship.

Thank you to Professors Stephen Hinshaw and Robert Knight in the Psychology Department at UC Berkeley, for taking time out of your busy days to meet with me to discuss stroke and the brain as it pertains to neuroscience. I knew my alma mater would come through for me with generous insight.

Thank you to my doctors and gurus. My cardiologist and neurologist, Dr. Neal White and Dr. Brad Volpi, who expertly navigated my body's healing. Tara Stiles at Strala Yoga in New York City—where I began my relationship with yoga and a healthy understanding of my body. And to my longtime therapist, Karl Knobler, who helped guide my mind's healing.

Thank you to the National Stroke Association and to the American Stroke Association; in those early months, they were a great resource to me.

I have been writing for years and years. This book is not the first book I've written, and so my writing mentors are many. Even if their eyes did not see this particular book in draft form, I owe so much to them—for their insights into craft, for their encouragement, and especially for developing rigor. This book would not exist without you. Thank you to all my teachers, who are also my friends: Junot Diaz, Mat Johnson, Victor LaValle, Chris Abani, Yiyun Li, Elmaz Abinader, Chang-rae Lee, and the late Justin Chin. And deep gratitude to VONA, for being my writing home and for nurturing my social consciousness as a writer.

My fambam spans both blood and love, and what I know about love comes from you: Mom; Dad; my brother, Richard; Penelope; and Orion Letizi.

Thank you to A. We had a good run while it lasted.

And to my daughter, Penelope: I am proud to be your mother.